Praise for *The Honey Bus*

"A moving memoir… A fascinating and hopeful book of family, bees, and how 'even when [children] are overwhelmed with despair, nature has special ways to keep them safe.'"
—*Kirkus Reviews*

"To read about Meredith May's bee family and her human family is to garner heart strength. A true story in every sense."
—Maxine Hong Kingston, bestselling author of *The Woman Warrior*

"An unforgettable story."
—HGTV.com

"[May's] prose is tender, thoughtful and transporting… [A] memoir of aching loneliness, reckoning and redemption. Beautiful and brave."
—Domenica Ruta, *New York Times* bestselling author of *With or Without You: A Memoir*

"A gripping narrative."
—*San Francisco Chronicle*

"Clear eyed, often very funny, and agonizingly compassionate."
—Laline Paull, author of *The Bees*

"[A] sharply visceral memoir."
—*Booklist*

"Everyone will leave this book with much more knowledge about bees and humanity, and the compassion that lives at the intersection of the two. [A] captivating coming-of-age family story."
—Noah Wilson-Rich, PhD, author of *The Bee: A Natural History*

The Honey Bus

A Memoir of Loss, Courage and a Girl Saved by Bees

Meredith May

ISBN-13: 978-0-7783-0975-8

The Honey Bus: A Memoir of Loss, Courage and a Girl Saved by Bees

This edition published by arrangement with Harlequin Books S.A.

Park Row Books
22 Adelaide St. West, 40th Floor
Toronto, Ontario M5H 4E3, Canada
ParkRowBooks.com
BookClubbish.com

Printed in U.S.A.

For Grandpa

E. Franklin Peace
1926–2015

The Honey Bus

"So work the honeybees, creatures that by a rule in nature teach the art of order to a peopled kingdom."

—William Shakespeare, *Henry V*

PROLOGUE

Swarm

1980

Swarm season always arrived by telephone. The red rotary phone jangled to life every spring with frantic callers reporting honeybees in their walls, or in their chimneys, or in their trees.

I was pouring Grandpa's honey over my corn bread when he came out of the kitchen with that sly smile that said we'd have to let our breakfast go cold again. I was ten, and had been catching swarms with him for almost half my life, so I knew what was coming next. He slugged back his coffee in one gulp and wiped his mustache with the back of his arm.

"Got us another one," he said.

This time the call came from the private tennis ranch about a mile away on Carmel Valley Road. As I climbed into the passenger seat of his rickety pickup, he tapped the gas pedal to coax it to life. The engine finally caught and we screeched out of the driveway, kicking up a spray of gravel behind us. He whizzed past the speed limit signs, which I knew from riding with Granny said to go twenty-five. We had to hurry to catch the swarm because the bees might get an idea to fly off somewhere else.

Grandpa careened into the tennis club and squealed to a stop near a cattle fence. He leaned his shoulder into his jammed door and creaked it open with a grunt. We stepped into a mini-cyclone of bees, a roaring inkblot in the sky, banking left and right like a flock of birds. My heart raced with them, frightened and awestruck at the same time. It seemed like the air was throbbing.

"Why are they doing that?" I shouted over the din.

Grandpa bent down on one knee and leaned toward my ear.

"The queen left the hive because it got too crowded inside," he explained. "The bees followed her because they can't live without her. She's the only bee in the colony that lays eggs."

I nodded to show Grandpa that I understood.

The swarm was now hovering near a buckeye tree. Every few seconds, a handful of bees darted out of the pack and disappeared into the leaves. I walked closer, and looked up to see that the bees were gathering on a branch into a ball about the size of an orange. More bees joined the cluster until it swelled to the size of a basketball, pulsating like a heart.

"The queen landed there," Grandpa said. "The bees are protecting her."

When the last few bees found their way to the group, the air became still again.

"Go wait for me back by the truck," Grandpa whispered.

I leaned against the front bumper, and watched as he climbed a stepladder until he was nose-to-nose with the bees. Dozens of them crawled up his bare arms as he sawed the branch with a hacksaw. Just then a groundskeeper started up a lawn mower, startling the bees and sending them back into the air in a panic. Their buzz rose to a piercing whine, and the bees gathered into a tighter, faster circle.

"Dammit all to hell!" I heard Grandpa cuss.

He called out to the groundskeeper, and the mower sputtered off. While Grandpa waited for the swarm to settle back down into the tree, I felt something crawling on my scalp. I reached up and touched fuzz, and then felt wings and tiny legs thrashing in my hair. I tossed my head to dislodge the bee, but it only became more tangled and distressed, its buzz rising to the high pitch of a dentist's drill. I took deep breaths to brace for what I knew was coming.

When the bee buried its stinger in my skin, the burn raced in a line from my scalp to my molars, making me clench my jaw. I frantically searched my hair again, and stifled a scream as I discovered another bee swimming in my hair, then another, my alarm radiating out wider and wider from behind my rib cage as I felt more fuzzy lumps than I could count, a small squadron of honeybees struggling with a terror equal to my own.

Then I smelled bananas—the scent bees emit to call for backup—and I knew that I was under attack. I felt another searing prick at my hairline followed by a sharp pierce behind my ear, and collapsed to my knees. I was fainting, or maybe I was praying. I thought that I might be dying. Within seconds, Grandpa had my head in his hands.

"Now try not to move," he said. "You've got about five more in here. I'll get them all out, but you might get stung again."

Another bee stabbed me. Each sting magnified the pain until it felt like my scalp was on fire, but I grabbed the truck tire and hung on.

"How many more?" I whispered.

"Just one," he said.

When it was all over, Grandpa took me into his arms. I rested my pounding head on his chest, which was muscled from a lifetime of lifting fifty-pound hive boxes full of honey. He gently placed his calloused hand on my neck.

"Your throat closing up?"

I showed him my biggest inhale and exhale. My lips felt oddly tingly.

"Why didn't you call out to me?" he asked.

I didn't have an answer. I didn't know.

My legs were shaky, and I let Grandpa carry me to the truck and place me on the bench seat. I'd been stung before, but never by this many bees at once, and Grandpa was worried that my body might go into shock. If my face swelled up, he said, I might have to go to the emergency room. I waited with instructions to honk the horn if I couldn't breathe as he finished sawing the branch. He shook the bees into a white wooden box and carried it to the truck bed while I reached up and checked the hot lumps on my scalp. They were tight and hard, and it seemed like they were getting bigger. I worried that pretty soon my whole head would be puffed out like a pumpkin.

Grandpa hustled back into the truck and started the engine.

"Just a minute," he said, taking my head in his hands and exploring my scalp with his fingers. I winced, certain he was pressing marbles into my head.

"Missed one," he said, drawing a dirty fingernail sideways across my scalp to remove the stinger. Grandpa always said that squeezing the stinger between your thumb and finger is the worst way to pull it out, because it pushes all the venom into you. He held out his palm to show me the stinger with the pinhead-sized venom sac still attached.

"It's still going," he said, pointing to the white organ flexing and pumping venom, oblivious that its services were no longer needed. It was gross, and made me think of a chicken running with its head cut off, and I wrinkled my nose at it. He flicked it out the window and then turned to me with a pleased look, like I had just shown him my report card with all A's.

"You were very brave. You didn't panic or nothin'."

My heart cartwheeled in my chest, proud of myself for letting the bees sting me without screaming like a girl.

Back home, Grandpa added the box of bees to his collection of a half dozen hives along the back fence. The swarm was ours now, and would settle into its new home soon. Already the bees were darting out of the entrance and flying in little circles to explore their surroundings, memorizing new landmarks. In a few days' time, they would be making honey.

As I watched Grandpa pour sugar water into a mason jar for them, I thought about what he had said about the bees following the queen because they can't live without her. Even bees needed their mother.

The bees at the tennis ranch attacked me because their queen had fled the hive. She was vulnerable, and they were trying to protect her. Crazy with worry, they'd lashed out at the nearest thing they could find—me.

Maybe that's why I hadn't screamed. Because I understood. Bees act like people sometimes—they have feelings and get scared about things. You can see this is true if you hold very still and watch the way they move, notice if they flow together softly like water, or if they run over the honeycomb, shaking like they are itchy all over. Bees need the warmth of family; alone, a single bee isn't likely to make it through the night. If their queen dies, worker bees will run frantically throughout the hive, searching for her. The colony dwindles, and the bees become dispirited and depressed, sluggishly wandering the hive instead of collecting nectar, killing time before it kills them.

I knew that gnawing need for a family. One day I had one; then it was gone overnight.

Not long before my fifth birthday, my parents divorced and I suddenly found myself on the opposite coast in California, squeezed into a bedroom with my mom and younger brother in my grandparents' tiny house. My mother slipped under the bedcovers and into a marathon melancholy, while my father was never mentioned again. In the empty hush that followed, I struggled to make sense of what had happened. As my list of life questions grew, I worried about who was going to explain things to me.

I began following Grandpa everywhere, climbing into his pickup in the mornings and going to work with him. Thus began my education in the bee yards of Big Sur, where I learned that a beehive revolved around one principle—the family. Grandpa taught me the hidden language of bees, how to interpret their movements and sounds, and to recognize the different scents they release to communicate with hive mates. His stories about the colony's Shakespearean plots to overthrow the queen and its hierarchy of job positions swept me away to a secret realm when my own became too difficult.

Over time, the more I discovered about the inner world of honeybees, the more sense I was able to make of the outer world of people. As my mother sank further into despair, my relationship with nature deepened. I learned how bees care for one another and work hard, how they make democratic decisions about where to forage and when to swarm, and how they plan for the future. Even their stings taught me how to be brave.

I gravitated toward bees because I sensed that the hive held ancient wisdom to teach me the things that my parents could not. It is from the honeybee, a species that has been surviving for the last 100 million years, that I learned how to persevere.

1

Flight Path

February 1975

I didn't see who threw it.

The pepper grinder flew end over end across the dinner table in a dreadful arc, landing on the kitchen floor in an explosion of skittering black BBs. Either my mother was trying to kill my father, or it was the other way around. With better aim it could have been possible, because it was one of those heavy mills made of dark wood, longer than my forearm.

If I had to guess, it was Mom. She couldn't stand the silence in her marriage anymore, so she got Dad's attention by hurling whatever was within reach. She ripped curtains from rods, chucked Matthew's baby blocks into walls and smashed dishes on the floor to make sure we knew she meant business. It was her way of refusing to become invisible. It worked. I learned to keep my back to the wall and my eyes on her at all times.

Tonight, her pent-up fury radiated off her body in waves, turning her alabaster skin a bright pink. A familiar dread pooled in my belly as I held my breath and studied the wallpaper pattern of ivy leaves winding around copper pots and rolling pins, terrified that the slightest sound from me would redirect the invisible white-hot beam between my parents and leave a puff of smoke where once a five-year-old girl used to be. I recognized this stillness before the storm, the momentary pause of utensils held aloft before the verbal car crash to come. Nobody moved, not even my two-year-old brother, frozen mid-Cheerio in his high chair. Dad calmly set down his fork and asked Mom if she planned to pick that mess up.

Mom dropped her paper napkin on top of her untouched dinner; we were eating American chop suey again—an economical mishmash of elbow macaroni, ground beef and whatever canned vegetables we had, mixed with tomato sauce. She lit a cigarette, long and slow, and then blew smoke in Dad's direction. I expected him to take his normal course of action, to unfold his long body from the chair, disappear into the living room and crank the Beatles so loud that he couldn't hear her. But tonight he just stayed seated, arms crossed, his coal-colored eyes boring at Mom through the smoke. She flicked her ash into her plate without breaking his stare. He watched her, disgust etched into his face.

"You promised to quit."

"Changed my mind," she said, inhaling so deeply I could hear the tobacco crackle.

Dad slapped the table and the silverware clattered. My brother startled, then his lower lip curled down and his breath hitched as he wound up for a full-body cry. Mom exhaled in Dad's direction again and narrowed her eyes. My nerves hopped like a bead of water in a frying pan as I nervously tapped my fingers on my thigh under the table, counting the seconds as I waited for one of them to pounce. When I counted to seven, I noticed the beginnings of a sardonic smile at the corners of Mom's mouth. She stubbed her cigarette out on her plate, rose and sidestepped the peppercorns, then stomped into the kitchen. I heard her banging pots, and then a lid clattered to the ground, ringing a few times before it settled on the floor. She was up to something, and that was never good.

Mom returned to the table with a steaming pot, still warm from the stove. She lifted it over her head and I screamed, worried she would burn Dad dead. He screeched his chair back, stood up and dared her to throw it. My stomach lurched, as if the table and chairs had suddenly lifted off the floor and spun me too fast like one of those carnival teacup rides.

I closed my eyes and wished for a time machine so I could go back to just last year, when my parents still talked to each other. If I could just pinpoint that moment right before everything went wrong, I could fix it somehow and prevent this day from ever happening. Maybe I'd show them the forgotten box of Kodachrome slides in the basement, the evidence that they loved each other once. When I first held the paper squares to the sunlight, I'd discovered that Mom's face was once full of laughter, and she used to wear short dresses and shiny white boots and smoke her cigarettes through a long stick like a movie star. She still had the same short boy haircut, but it was a brighter shade of red then, and her eyes seemed more emerald. In every slide Mom was smiling or winking over her shoulder at Dad. He took the photos not long after he'd spotted her registering for classes at Monterey Peninsula College, and invited her for a drive down the coast to Big Sur.

He'd recognized her from a few summer parties. She had been the one with the loud laugh, the funny one with a natural audience always in tow. He noticed how easily she flowed in a crowd of strangers, which drew my quiet father out of the corners. He was raised never to speak unless spoken to, and liked to study people before deciding to talk to them. This made him slightly mysterious to my mother, who was drawn by the challenge of getting the tall stranger with the dramatic widow's peak and smoky eyes to open up. When he told her his plan to join the navy and travel abroad after college, Mom, who had never been outside California, was sold.

They married in 1966, and within four years the navy relocated them to Newport, Rhode Island, where Matthew and I were born. After his service, Dad worked as an electrical engineer, making machines that calibrated other machines. Mom took us on strolls to the butcher and the grocery store, and made sure dinner was on the table at five. On the outside, our lives seemed neat, organized, on track. We lived in a wood-shingled row apartment, and my brother and I had our own rooms on the second floor, connected by a trail of Lincoln Logs and Lite-Brite pins and gobs of Play-Doh dropped where we'd last used them. Dad installed a swing on the front porch, and we played with the neighbor kids who lived in the three identical homes attached to ours. On weekend mornings, Dad came into my room and we identified clouds as they passed my bedroom window, pointing out the dinosaurs and mushrooms and flying saucers. Before going to sleep, he read to me from *Grimm's Fairy Tales*, and even though every story ended in a violent death of some kind, he never said I was too little to hear such things.

It seemed like we were happy, but my parents' marriage was already curdling.

I imagine they tried at first to manage their squabbles, but eventually their disagreements multiplied and spread like a cancer until they had trapped themselves inside one big argument. Now Mom's shouting routinely pierced the walls we shared with the neighbors, so their problems had undoubtedly become public.

I opened my eyes and saw Mom standing there in position, ready to throw the pot of American chop suey. Their threats arrowed back and forth, back and forth, his restrained monotone mixing with her rising falsetto until their words blended into a high-pitched ringing in my ears. I tried to make it go away by softly humming "Yellow Submarine." It's the song Dad and I sang together with wooden spoons as our microphones. Back when music filled our house. Dad recorded every Beatles song off the radio or vinyl records onto spools of tape, which he kept in bone-colored plastic cases on the bookshelf, lined up like teeth. He listened to tapes on his reel-to-reel player, and lately he preferred "Maxwell's Silver Hammer," the one about the man who bludgeoned his enemies to death, blasting it from the living room until Mom inevitably told him to turn that racket down.

I was somewhere in the second verse when I saw her lift her arm, and the pot handle released from her palm seemingly in slow motion. Dad ducked, and our leftover dinner flashed through the air and slapped into the wall, where it slid down, leaving a slick behind as it pooled with the peppercorns on the floor. Dad picked the pot up from near his foot and stood, his whole body quivering with silent rage. He dropped the pot onto the table with a loud thud, not even bothering to put it on a hot plate like he was supposed to. Matthew was wailing now, lifting his arms to be picked up, and Mom went to him, as if nothing had just happened. She bounced Matthew, shushing softly into his ear, her back to Dad and me. Dad turned on his heel and escaped to the attic, where he would spend the night tapping out Morse code on his ham radio in conversation with polite strangers.

I didn't bother asking permission to leave the dinner table. I made a run for the staircase, two-stepped it up to my room and slammed the door. I pulled my Flintstones bedspread off and dragged it under my bouncy horse. It was a plastic horse held aloft by four coiled springs—one on each leg attached to a metal frame. I put my feet under its felt belly, and pushed it up and down until I'd established a soothing rhythm. I curtained my eyes with my shoulder-length hair, blotting out reality so that I could almost believe that I was safe inside a yellow submarine, below the surface, alone, and so far down I couldn't hear any voices at all.

Although I didn't understand why my parents fought so much, deep down I understood that something significant was shifting inside our house. Dad had stopped using his words, and Mom had started using too many. I tried to make sense of it by gleaning bits of information I overheard whenever my godmother, Betty, dropped by while Dad was at work. Mom and Betty would sit on the couch and talk about all sorts of things while Betty would play with my hair. Matthew would go down for his nap, and I'd sit on the carpet between their legs where Betty could reach down and absentmindedly wind long strands of my brown hair around her fingers. She'd twist my locks into knotted snakes and then let it unfurl, over and over, while she and Mom worked out their problems. She'd coil my hair tight, then release. Twist, tug, release. Twist, tug, release. It felt like getting a deep itch scratched, a tingling scalp massage that could go on as long as it took them to smoke a whole pack of cigarettes.

They talked the afternoons away, and I stayed so quiet that they forgot about me and got to discussing things I probably shouldn't have heard. Mostly I learned that men are disappointing. That they promise the moon, but then don't bring home enough money for groceries. I overheard Mom say that Dad might lose his job because his boss was doing something called "downsizing."

"Layoffs?" Betty asked. *Twist, tug, twist, tug.*

"Apparently," Mom said. "They're letting all the junior engineers go."

"Shit on a shingle."

"You said it."

"What will you do?" *Twist, tug.*

"Hell if I know."

Betty tugged on my hair once more and let it uncoil from her index finger. I stayed statue quiet, ear hustling. They were silent for a few minutes, and Betty switched to scratching my scalp, sending pollywogs of ecstasy squiggling down my neck. Mom got up and fetched two more Tab sodas from the fridge and cracked them open, handing one to Betty. Mom plunked back down onto the sofa and put her feet up on the sagging ottoman. She sighed so hard it sounded like she was deflating.

"Honestly, Betty, I don't think marriage is all it's cracked up to be. I'm thirty and feel like ninety."

Betty shifted her heavy legs, unsticking them from the Naugahyde and stretching them out lengthwise. She attempted a forward bend, but couldn't reach her hands much past her knees. She grunted with effort and sat back up. She pushed aside the curtains and looked out the window.

"You think being single is all rainbows and unicorns?"

Mom blew a wedge of smoke out one side of her mouth and dropped her stub into an empty pink soda can where it hissed out. "At the rate this is going," Mom said, "I'd be happy to change places."

Betty turned back and looked directly at Mom, to make sure she had her full attention. "Sometimes it's lonely."

"It's better to be lonely alone than lonely married."

Betty cocked an eyebrow at Mom as if to say she wanted proof. Mom launched into Exhibit A—the time she was returning from a walk with me in the buggy, and Dad hollered down to her from the upstairs window to come quick. Terrified something was wrong with Matthew, she left me in the buggy on the sidewalk and streaked into the house and up the stairs, only to find the crisis was a diaper that needed changing.

Mom's voice turned indignant. "*Isn't* child rearing supposed to be fifty-fifty?"

Betty let out a low commiserating whistle. I wanted to ask if Mom ever went back outside for me in the buggy, but knew it wasn't the time to remind them I was listening.

"Betty, listen to me. Don't marry anyone without first asking one crucial question."

Betty's fingers froze in my hair temporarily, waiting for the secret to marital bliss.

"Ask if he's willing to change diapers. Depending on his answer, he'll treat you as his equal, or his employee."

I lifted my head like a cat to prod Betty's fingertips and remind her of her job. Her fingers automatically hooked a strand of my hair and began winding it into a knot. I knew that I was not to repeat anything that was said on the couch. It made me feel a little squirmy to eavesdrop on them, but I liked the head scratching too much to pull myself away.

I must have fallen asleep under the bouncy horse, because I didn't remember how I got into bed when Mom pushed open my bedroom door with such force it slammed into the wall, jarring me awake. She yanked open dresser drawers, and tossed fistfuls of my clothing into a white suitcase with satiny orange lining. I sat up and tried to adjust my focus, but she was moving so fast she stayed blurry.

"Five minutes," she said, standing still for a second. "I'm going to get your brother. Be dressed by the time I get back."

Mom whizzed out of my room. It was dark outside. My

body felt like concrete, and I didn't want to go out into the cold. Mom had done this before. She'd shake us awake in the middle of the night, hurry us into snow pants and hats and mittens, and run down the stairs screaming that she was going to run away. Dad would let her scurry around the house packing until she tired herself out, then he'd eventually get her to sit next to him on the couch to talk. He had a low soothing voice, and she was like a too-loud TV. From the top of the stairs, I'd listen until there was no more yelling and I heard her sniffling, the signal that the argument had passed and it was time for everybody to go back to sleep.

I decided to wait Mom out this time. When she reappeared in my door frame with Matthew on her hip, I was still sitting like a question mark in bed.

"Where are we going?"

"Not now, Meredith. I'm in *no* mood."

Balancing my brother in one arm, she tugged off my pajamas and wrestled me into daytime clothes. Mom was scooting me toward the door when I turned back.

"Can I bring Morris?"

Morris was a stuffed pink cat with a skirt that my parents had bought at a drugstore on the way home from the navy hospital nursery after I was born. I had named him Morris after the cat in the TV commercial, and he was my most prized possession. I had grown so dependent on him, especially lately, that I couldn't fall asleep if he wasn't tucked under my arm. Mom nodded her permission, and I dug around my sheets, grabbing him just seconds before Mom led me out of the room by my wrist.

As Mom was helping me into my coat in the hallway, Dad passed by, his shoulders slumped in defeat. He opened the front door and walked out into the chilly air. I ran to the living room window and watched as he started up the Volvo under the light of the porch. His breath came in silver puffs as he scraped frost from the windshield. I watched him lift the suitcase into the trunk and get into the driver's seat while Mom strapped Matthew in the car seat and then came back inside for me. I clutched Morris closer to my chest, and rubbed my chin back and forth against the soft fleece of his pink ears.

"Where are we going?" I asked again, softer this time. Mom zipped up my puffy jacket and put her hands on my shoulders.

"California. To visit Granny and Grandpa."

Her voice warbled, but she forced a smile and I brightened just a bit. Last summer Granny and Grandpa came for a visit, and because they were guests there was no fighting in our house for a whole week. Grandpa and Dad took me to the beach and taught me how to bodysurf, letting the waves lift and slingshot me into the hissing foam until I glided to a stop on my belly in the sand. Grandpa put me on his shoulders and dug quahogs out of the mud with his toes, teaching me how to spot spurts of water where the clams were siphoning. We brought home a whole bucket and shucked them in the kitchen for dinner. Maybe there'd be quahogs in California.

Inside the car, Mom turned away from Dad and drew wet lines on the frosty window with her finger. Matthew fell back asleep with his head bent toward me, his light brown hair falling into his eyes and his little red lips making a puff noise instead of an actual snore. Unlike me, who came into the world screaming, my brother arrived, blinked twice and smiled. Mom liked to say that I had apparently used up all the fussy and left none for him. It was true; Matthew's soul was calm and trusting. He was a boy who assumed goodness in everyone. What three-year-old smiled while you took candy out of his hand, certain the game would end with something even better in return? I could feel Matthew's trust in humankind when he curled his hand around my index finger and toddled in a tipsy lockstep with me, certain I wouldn't let him fall. He followed me everywhere, plucking words out of my sentences and parroting them like my own personal backup singer. It was for those kinds of things that I loved him fiercely, even though he wasn't much of a conversationalist. But he knew one word that bonded me to him for life. Whenever he awoke from a nap and saw me walk into his room, he'd stand and reach for me with chubby starfish hands.

"Mare-miss!" he'd shout.

I had a super fan, and his adoration gave me a profound sense of distinction.

Dad shifted gears with punching force, and I hugged my knees to my chest and rocked in the back seat, silently willing someone to speak. Mom spoke just once on the ninety-minute drive to the airport in Boston; she asked Dad to detour to Fall River so she could stop at a friend's house to say a quick goodbye. When we finally pulled into the airport parking lot, suddenly everything was moving too fast. Doors opened and slammed. The four of us speed-walked in silence. As the glass panels of the revolving door spun around us, I felt like I was falling down a well. I didn't understand what was happening, other than it was big, and that I wasn't supposed to ask about it. I grabbed Mom's hand and held on.

Dad bought our tickets and handed our suitcase to the woman behind the counter, and I watched it glide away on a conveyor belt and disappear through an opening in the wall. When we reached the gate, Dad brought me to the window and pointed out the plane we were going to take to visit Granny and Grandpa. It gleamed in the morning light, a sleek bird with upturned wings, and I felt a flutter inside, imagining myself soaring inside it. I peppered Dad with questions—how high would the plane go, how did it stay in the air, would he sit next to me? When it was time to board, Dad knelt down and squeezed me so hard that I felt him shaking.

"You be good, kiddo," he said, forcing a smile. "Love you."

My body suddenly turned cold. I felt something rip inside my stomach as Dad sank into an airport chair and Mom tugged me toward the door leading to the plane. This wasn't right. Dad was supposed to come with us. Mom pulled me by the arm as I leaned in the opposite direction, unwilling to take another step without Dad.

"Come ON," she huffed.

"What about Dad?" I demanded, digging in my heels. But she was stronger, and I was forced to hop in her direction as I struggled against her weight.

"Don't make a scene."

I let myself go slack. Conversation around me became muffled, like I was underwater. I fell silent, feeling myself get pulled into the breezeway, and when I looked back to find Dad, there were too many people behind me, blocking my view. My mind swirled as I let Mom steer me down the aisle and into a window seat, where I pressed my forehead to the chilly oval until I saw a tall figure with ink-black hair and plaid pants standing behind the plate glass of the terminal. Dad looked like he was in a television. I lifted my hand, but he didn't see me. He didn't move from his spot as the plane pushed back from the gate. I kept my eyes locked on him until he became smaller and smaller, until the plane turned away.

During the flight, Mom blew smoke at the folding tray in front of her and picked at her copper-colored nail polish with trembling hands. She seemed to be crumbling. I snuck peeks at her while pretending to draw in the coloring book the stewardess had given me. Mom still looked pretty to me, but her skin seemed grayer under the overhead light. At home, she was careful about the way she looked, and never went outside without first covering her freckles with beige cream and putting shimmery blue shadow on her eyes. I liked to watch her ritual, and all the tools that came with it. A blow-dryer to make her short curly hair stand up higher, fat brushes to put pink powder on her cheeks, and that clamper thing she squeezed on her eyelashes to curl them up. Sometimes she'd let me choose her lipstick from dozens of tubes she kept in the bathroom. The final touch was a cloud of smelly spray all around her head, to make her hair stay in place.

"It doesn't matter if you're a little chubby, as long as you have a pretty face," she'd say, threading gold wire hoops through her ears. She never left the house without her movie-star sunglasses, two big brown circles as large as drink coasters.

Mom had some rolls around her middle but her legs were thin, so she covered her shape with dresses that had busy designs and loud colors. The dresses stopped above her knee, which made her look like a bouquet of flowers on two stems. I thought she was beautiful. My favorite part of watching her get dressed was when she picked out her shoes. She kept a row of heels in a perfect line on her closet floor, toes facing in, in every color of the rainbow. I wasn't allowed to touch her things, but I admired her footwear, imagining myself perched high like a lady, strutting down the sidewalk to my grown-up job. Once she'd put on her outfit, she'd turn left and right in the mirror and ask me if she looked fat. I never thought so, but she always looked disappointed when she looked at her reflection.

At least once a month, she got dressed up to visit the Vanderbilt mansion. The towering limestone "summer cottage" had seventy rooms and looked like six houses pushed together, perched on a bluff overlooking the Atlantic. It was a five-minute drive from our apartment, and we entered through the wrought-iron gates, Mom's dress rustling softly and Charlie perfume wafting behind her, as she pushed Matthew in the stroller past topiaries clipped to scientifically precise triangles, the pea gravel pathway crunching underfoot. We never went inside for the tour, but we had our favorite bench where Mom had a view of the top floor windows. My brother picked pebbles out for me to throw into the garden fountains as she conducted surveillance on the windows, hoping for a glimpse of one of the heirs who reportedly lived in the attic apartment.

Mom was absolutely engrossed during her mansion visits, as if familiarizing herself with opulence so that she'd be ready when prosperity came for her. She read books with Pygmalion plots about regular people being plucked from obscurity for greatness, gravitated toward movies about unearthing hidden treasure, and game shows of all kind. Mom was a dreamer without a plan, and as the years piled up without her Cinderella transformation, she felt more and more cheated out of the grandeur she was entitled to, and increasingly disappointed in my father for not providing it. She was forever waiting for life to happen to her and becoming more befuddled as to why it was not.

The plane made a little hop as it encountered some weather, and I snuck another glance Mom's way. She appeared drowsy, her eyes open but no expression behind them. Wadded Kleenex collected in her lap, and black makeup ran down her cheeks, smudged in places where she'd tried to wipe it away so it looked like bruises. Every once in a while she gave out a long, body-slumping sigh that sounded like all the air was coming out of her. I patted her arm, and she put her hand over mine absentmindedly. I wanted to ask why Dad wasn't coming with us, but knew I wouldn't get an answer. Even though her body was in the chair next to mine, her mind was somewhere else. I flipped the metal cover of the ashtray embedded in the armrest—open, closed, open, closed—hoping the noise would become so irritating that she'd have to talk, to tell me to knock it off.

If only Mom would say something. I wanted her to cry, or shout, or throw something to send me a signal that things were still the same. But she was eerily quiet, and that was terrifying. At least with an outburst, I could tell what was on her mind. Silence was not her style, so that meant something serious was happening. Dread dripped in the back of my throat, an acrid taste like burned walnuts.

I tried to keep a vigil over her, but eventually the engine hum inside the cabin lulled me to sleep. I dreamed there was a small reservoir in the floor of the plane near my feet, with a long lever protruding from it. I unfastened Matthew's seat belt and shoved him into the hole and pulled the lever. Hissing steam rose, and when I released my grip, Matthew had turned into a blue glass totem, about the size of a soda can. He was trapped in the glass, and I could hear him screaming to be let out. I shoved him into my pocket, promising him that I would turn him back into a boy, but for now, this was the best way to keep him safe until we arrived at Granny and Grandpa's house.

My intuition was telling me that I needed to protect my little brother. During the flight, I could sense that Mom was receding from us. I felt a slipping away that I couldn't put words to, a change as subtle as growing taller that couldn't be perceived until it had already happened. By the time we landed, her eyes were vacant and looked right through me. Somewhere thirty thousand feet up over Middle America, she had relinquished parenthood.

2

Honey Bus

Next Day—1975

Granny was waiting for us at the Monterey Peninsula Airport, standing with arms crossed in a wool dress and a crisp, high-collared blouse with puffy sleeves. Her tawny bouffant was salon-sculpted into frozen waves, and protected by a clear plastic headscarf tied under her chin to shield her hairdo from the elements. She was an exclamation point of perfect posture, jutting above the glut of less-mannered travelers flagrantly kissing their relatives in public. She scrutinized our approach through cat-eye glasses, lips pursed in a thin line. When Mom saw her, she let out a wounded cry, and reached for a hug just as Granny pulled out a wadded hanky from her sleeve cuff and held it out to Mom to avoid an embarrassing scene. Mom took it and just stood there, unsure of what to do. Granny observed manners, and one did not blubber in public.

"Let's have a seat," Granny whispered, grabbing Mom's

elbow and guiding her to the row of hard plastic chairs. Mom blew her nose and gulped back sobs as Granny made soft clucking noises and rubbed her back. I stood there awkwardly, looking while at the same time trying not to look. Granny handed Matthew and me two quarters from her coin purse and pointed to a row of chairs with small black-and-white televisions mounted on the armrests. Delighted, we ran to the chairs to watch a TV show while Mom and Granny had a Very Important Conversation. Matthew and I squeezed together in one of the chairs, dropped the quarter in and spun the dial until we landed on a cartoon.

When Granny and Mom finally stood up to go, we were the last people left in the boarding area. Granny came over, and I instinctively stopped slouching. "Your mother is just tired," she said, leaning down to kiss my cheek. She smelled like lavender soap.

Matthew and I rode in the wayback of Granny's mustard-yellow station wagon, far enough from Granny and Mom so we couldn't hear what they were saying. I looked out the back window to inspect California sliding past. It was February, but oddly there wasn't any snow. We drove over rolling brown hills with horse ranches and up a steep grade with hairpin turns, pushing the car higher and higher. The car groaned with effort, and my stomach dropped when I realized that we were on top of a ring of mountains, like we were driving on the edge of a gargantuan bowl. Beneath us, the earth fell away in deep folds and grooves all the way to the valley below, and an idea came to me that we must be driving over the dinosaurs, whose bodies had turned into mountains after they'd died.

I also noticed that the trees in California were different—solitary, massive oaks with outstretched octopus arms twisting just a few feet above the ground, nothing at all like the fiery maples or crowded forests of skinny birch trees back home. When we finally started to descend, I could see all of Carmel Valley below us, a vast green basin with a silver river snaking along one side of it. My ears popped on the way down until we reached the bottom of the bowl, the mountains now a towering fortress around us. Carmel Valley felt like a secret garden in one of my fairy tales, sealed off from the rest of the universe. It was warmer here, and the sun seemed to slow everything down: the ambling pickup trucks, the sleepy crows, the unhurried river.

We drove by a community park and public swimming pool, then made a right turn onto Via Contenta and passed an elementary school with tennis courts. The rest of the residential street was lined with one-story ranch homes separated by juniper hedges and oak trees for privacy. Granny slowed in front of a volunteer fire station where some men were washing red engines out front, passed a small cul-de-sac with a handful of identical wood-shingled bungalows, and then reached her destination—a small red home perched in the middle of an acre of land, bordered on four sides by overgrown trees.

Granny skipped her front gravel driveway and instead took the back way to the house, turning onto a short dirt lane that ran along her fence and was canopied by a row of mammoth walnut trees with branches reaching all the way to the ground, engulfing us in a tunnel of green leaves. Walnut shells popped under our tires as we followed the curving drive to the backyard. She parked next to a clothesline, where her square-dancing petticoats were flapping in the breeze.

Granny took great pride in living on one of the largest lots on her street, and she was quick to remind anyone who forgot that she was among the first residents of Carmel Valley Village, arriving in 1931 from Pennsylvania with her mother when she was eight. They'd driven across the country in a convertible Nash Coupe after Granny's father had unexpectedly died of a heart attack, because her mother wanted to escape the tragedy in a warmer place with good swimming. This history, Granny believed, conferred on her a pedigree that allowed her to complain about the influx of newcomers over the next forty years. However, she was comforted that the oak, walnut and eucalyptus trees demarcating her property had grown to screen the neighbors from view. And the neighbors in turn were spared the sight of Grandpa's accumulating junk heaps that now pervaded the king-size lot.

I stepped out of the car and saw several haystack-size piles of tree trimmings, at least three toolsheds, mounds of gravel and bricks, two rusting military jeeps, a flatbed trailer, a backhoe and two beaten-down pickups. A trellis of grapevines led in a sloping line from the laundry to the back fence, where there was a small city of stacking beehives resting on cinder blocks, each one four and five wooden boxes high. From this far away, it looked like a mini-metropolis of white filing cabinets.

Something caught my eye through the billowing laundry. I walked through the rainbow of swirling skirts to get closer, and found myself standing before a faded green military bus. Rain had chewed away a ring of rust holes around the roof, leaving brown streaks trailing down its sides. Weeds choked the tires, its wraparound front windshield was cracked and cloudy, and a massive rhubarb bush sprouted from under the front bumper. It seemed to have driven right out of World War II and wheezed to a stop right by Grandpa's vegetable garden, from an era when vehicles were all fat curves instead of sleek edges, making the bus look more animal than machine. The rounded hood was sculpted like the snout of a lion, with vent holes for nostrils and globe headlight eyes that stared back at me. Below its nose was a row of grinning grille teeth, and under that, a dented metal bumper that looked an awful lot like a lower lip. In peeling white paint above the windshield, it read U.S. ARMY 20930527. Captivated by the incongruity of it, I felt compelled to investigate.

Kicking a path through waist-high weeds, I tried to see inside but the windows were too high. I circled to the back of the bus, and near the tailpipe I found a crooked stack of wooden pallets that improvised as stairs leading to a narrow door. I scrambled up, the makeshift staircase wobbling beneath me, and pressed my nose to the filmy glass.

Inside, all the seats were gone, and in their place was some sort of factory of whirligigs, crankshaft gears and pipes. A metal basin about the size of a hot tub rested on the floor, and contained a hefty flywheel powered by pulleys as large as manhole covers. Behind the driver's seat were two massive steel barrels with cheesecloth stretched across their open tops. An overhead network of galvanized steel pipes was suspended from the ceiling with fishing lines.

The equipment ran the length of one wall, and on the other side Grandpa had stacked a bunch of wooden boxes, each about six inches tall and two feet wide, and painted white. Each rectangular box, taken straight from his hives, was open on the top and bottom and contained ten removable wood-framed sheets of wax honeycomb. The frames hung in neat rows, supported by notches inside the box. I would later learn from Grandpa that these were the "honey supers," the removable top-tier boxes of a modular beehive where the bees stored nectar in the wax honeycomb and thickened it into honey by fanning their wings. The supers rested atop the larger brood boxes at the base of the hive where the queen lays her eggs.

There must have been three dozen boxes of honeycomb inside the bus. Glistening honey trickled down the stacks, collecting in shiny pools on the black rubber floor.

I could see glass jars on the dashboard that had turned purple in the sun, and sunflower-yellow bricks of beeswax that Grandpa had made by melting wax honeycomb and straining it through pantyhose into bread pans to harden. Electrical cords snaked everywhere, and construction lights dangled from the ceiling handrails. I cupped my hands over my eyes to shield the glare, and from out of the shadows someone inside pressed their nose to mine. I startled and nearly fell backward, just as Grandpa popped out the back door.

"Boo!" he said.

Bees buzzed around his head, and he slammed the door

quickly to keep them from getting inside the bus. He was wearing threadbare Levi's a couple inches too short and no shirt. He had Einstein hair sticking out every which way, as if electricity had just zapped through it, and a round face tanned to a chestnut color that settled into an expression of bemusement with life, as if he was forever chuckling at a private joke. In one hand he held a can with smoke pouring out of a spout on top. He yanked a tuft of green grass out of the ground, jammed it into the spout to stifle the flame and set his bee smoker on a pile of bricks. Then he dropped down on one knee and opened his arms wide, signaling me to fall into them.

"I've been waiting for you," he said, squeezing me tight.

I peeled my arms from Grandpa's neck and pointed at the bus.

"Can I go in?"

His workshop held a Willy Wonka–like spell over me. He'd built it himself, out of hand-me-down beekeeping equipment and spare plumbing parts, and powered it with a gas-powered motor taken from a lawn mower. When he bottled honey inside during the hottest days of summer, the whole bus rumbled as if it were about to drive off, and the indoor temperature shot above one hundred. Nothing in his secret workshop was official, or safety-checked, and the sweltering, sticky danger of it all made entry that much more irresistible. It seemed like magic to me that Grandpa brought honey supers inside, and emerged hours later with jars of golden honey that tasted like sunlight. Grandpa had the power to harness nature, like Zeus, and I wanted him to teach me how.

Grandpa stood up and blew his nose into a grease-stained rag, then shoved it in his back pocket.

"My honey bus? It's not a place for little kids," he said. "Maybe when you're fifty, like me." The bus was too hot and dangerous inside, he said. I could lose a finger.

Grandpa reached his long arm to the roof of the bus, where he'd stashed a piece of rebar that was bent at a right angle. He inserted one end of the rod into a hole where the back door handle used to be, and twisted to lock the bus. Then he put the homemade key back on top of the bus, out of my reach.

"Franklin, would you come get the suitcase!" Granny called out, in a way that sounded more like an order than a question. Granny had honed her leadership skills with decades of practice keeping elementary schoolchildren in line. I was a little afraid of her, and always tried to be on good behavior because her presence inherently demanded it. Not only of me, but of everyone in her orbit. Grandpa's ears perked at the sound of her voice.

I followed Grandpa to the station wagon. He fetched our one shared suitcase from the back and we walked to the front door, trailed by a handful of bees attracted to the honey stuck to Grandpa's boots.

My grandparents lived in a tiny red house with a flat, white gravel roof that looked like year-round snow. Grandpa said it turned away the sun and was cheaper than air-conditioning. The house had two bedrooms, and a kitchen wrapped by an L-shaped room with redwood paneling that served as both the living and dining room. A large brick fireplace that took up half of one wall was the main source of heat. Next to it was a windup grandfather clock, and on the opposite side of the house, floor-to-ceiling windows facing the Santa Lucia Mountains, which formed a natural barrier between our house and Big Sur on the other side. The kitchen was painted baby blue and was home to Grandpa's black dachshund, Rita, who slept under the stool next to the washing machine. There was one bathroom, decorated with brown-and-silver-striped wallpaper, and a low-flow showerhead that misted weakly.

Granny led us to the spare bedroom that used to be Mom's when she was a girl. It had since been painted a cantaloupe color. I stepped inside and immediately saw my world shrink: Matthew would sleep in a cot in the corner, and I would share the double bed with Mom. We would put our clothes into a Victorian marble-topped vanity with two lavender-scented drawers. My room in Rhode Island suddenly seemed like a castle in comparison to this small box, so crowded by beds there was no space to play.

Mom immediately closed the curtains to the sun, sending a shadow over the walls. Granny steered Matthew and me back into the hallway.

"Your mother needs some peace and quiet," she whispered. "Go on and play outside."

Granny had a voice that never suggested, but always instructed. We immediately understood the first unspoken rule of our new home—Granny was in charge. She would be the one to set our daily routine, plan the meals and make decisions for Mom, Grandpa, us.

Mom didn't join us for dinner that night, so Granny put a bowl of tomato soup and toast on a tray for her instead. She set a crystal vase with a rose next to the bowl, like hotel service.

"Someone get the door," Granny said, standing before Mom's bedroom.

I twisted the doorknob and pushed, sending a wedge of yellow light into the darkened room, and a plume of cigarette smoke billowed out. The air was so thick I could feel it pour into my lungs as I inhaled. I took a step back and let Granny go in first. She gently approached the bed, where Mom was curled in a fetal position, crying softly. A glass ashtray the color of amber rested on the headboard, filled with a cone of ash.

"Sally?"

Mom moaned by way of answer.

"You should eat something."

Mom uncurled herself and sat up. She winced and squeezed her temples.

"Migraine," she whispered. Her voice was so thin, it sounded like it might tear. Granny flicked on the light, and I could see Mom's face was flushed and her eyes were puffy.

"Tylenol?" Granny offered, fishing the plastic bottle out of her pocket and rattling it.

Mom extended her arm, and Granny dropped two pills in her palm. Granny held out a water glass and Mom gulped twice, handed it back, then flopped back down into the pillows.

"The light," she said.

I reached up and turned it back off.

Mom seemed so weak, like she couldn't even hold her head up. I thought of that time I found a baby bird that had fallen out of its nest. It was pink, and I could see the blue of its bulging eyes that had yet to open. The poor thing's head lolled to the side when I tried to pick it up.

"I'll just leave this here," Granny said, setting the tray at the foot of the bed. Mom waved it away. Granny stood over the bed for a few seconds, waiting for Mom to change her mind. She bent down and adjusted the pillows to make Mom more comfortable, then Mom closed her eyes again and turned away from us. Granny picked up the tray and we shuffled out.

That first night Matthew slept in his new cot, while I crawled into the big bed where Mom was burrowed into the middle, the sheets tightly wound around her like a burrito. I carefully tugged on the sheet, trying not to wake her. She mumbled in her sleep and half-heartedly tugged back, then scooted aside to make room for me. She sniffled and fell into a light snore.

I moved to the edge of the mattress, as far as I could possibly be from Mom without falling out of the bed. I faced the window, which ran the length of the wall, tracing the moonlight that leaked in around the perimeter of the curtains with my finger. I didn't want our bodies to touch, as if her tears were contagious.

I felt twitchy and sleep wouldn't come. I wondered what Dad was doing at that moment, if he was walking through the empty rooms of our house, changing his mind and deciding to come to California after all. I hoped that whatever had just happened to our family was temporary, but I didn't understand what had broken, so I couldn't imagine how to fix it. I had a new uneasiness in the pit of my stomach because I now knew the injustice of random bad luck, that it was possible to have a family one day and lose it the next. I wanted to know why I was being singled out for punishment, and tried to retrace my steps to pinpoint what I had done wrong to have my life upended this way. It was baffling, but I had the sense that going forward, I had to choose my words, and my steps, more carefully, so I could do my part to comfort my mother and slowly, craftily, coax her happiness back. I had to be good, and patient, and maybe my luck would turn around.

Mom's and Matthew's snores settled into a syncopated rhythm, and I tried to match my breathing to theirs so I could relax into sleep. I lay motionless and fell into a self-induced trance, humming "Yellow Submarine" quietly until I receded somewhere deep inside my skull and blinked out.

Over the next few weeks, Mom remained bedridden. Granny tried various strategies to cheer her up and brought her all sorts of bedside meals, trying to find something she could stomach. But Mom refused most of it, accepting only sugary coffee, canned soda and the occasional bowl of cottage cheese. Granny fetched hot pads for her back, cold compresses for her forehead and murder mysteries from the library. Still Mom's migraines wouldn't go away. When she complained of sore muscles, Granny dug around in the hall closet and produced a gadget that looked like a handheld electric mixer; only this thing had one stem protruding from it that ended in a flat metal disc. Granny plugged it in and the disc heated up and vibrated. She sat on the bed and moved the vibrator across Mom's back in lazy arcs, loosening the tension while Mom sighed with relief.

My brother and I were not allowed in the bedroom during daytime because Mom needed to recuperate, but Granny would sit at her bedside for hours in deep conversation, and despite my eavesdropping I caught only snatches of it. Mostly I heard Granny reassuring Mom that it wasn't her fault, that she could put this behind her, that men were worthless when you really get right down to it, and not worth this much fussing over. I'd hear Mom sniffling and asking wounded questions. Why me? What am I supposed to do now? What did I ever do to deserve this? Her questions were similar to my own, and I strained to hear an answer from Granny that would explain. One never came, and eventually I grew weary of spying and gave up.

Spring came, and the almond tree in the front yard erupted in white flowers. Mom entered her third month of bed rest, yet her despondency only grew. Mom's bad luck invoked in Granny an inexhaustible pity. While Granny gave Mom a safe haven and unlimited time to regain her strength, she worked double-time to hold up appearances that my younger brother and I weren't really semiorphans. She never spoke to us about what was happening to our mother, instead forging ahead as if nothing was amiss. Granny bought and washed our clothes, took us to the doctor for checkups, made us brush our teeth before bed and wrote scathing letters to our father demanding he send more money to support us. Granny adapted to her second motherhood with a sense of family duty, which gave Mom permission to carve out a new identity as a woman scorned. Granny looked after Matthew and me in an obligatory way, without the affection she reserved for her daughter. Mom was her child, and we were more like unexpected foster kids. In her most frustrated moments, she blamed Matthew and me for ruining her life plans, letting us know that if it weren't for that no-good father of ours, she could have been enjoying her retirement.

Her suggestion to go outside and play became a refrain. Granny now had more laundry to do, more food to make, more dirt tracked in the house to clean up and she couldn't keep up with it if we were constantly underfoot.

Outside there was plenty to mess with, and as we were loosely supervised by our grandparents, we were free to roam the yard as long as I kept an eye on my little brother. That first summer Matthew and I gorged on Grandpa's blackberry vines until our lips and fingertips were purple. We climbed into two hollowed-out army jeeps rusting in the yard and drove them through dozens of imaginary wars. We unearthed plastic soldiers and old glass marbles that someone had buried in "olden times," and we came upon an enormous pruning pile that Grandpa had been contributing to since before we were born—a colossal hill of fruit tree branches—and scaled it on all fours like lizards climbing a wall. We discovered that if we jumped up and down on the heap, we got excellent bounce, just like a trampoline. We fell off and bruised ourselves only a few times.

We quickly adjusted to the outdoor sounds of Carmel Valley, no longer jumping in terror when one of the peacocks on the hilltops let out a squeal like a woman being throttled, and learned to differentiate between the ambulance and fire sirens coming from the volunteer fire station down the block. We much preferred outside to inside, which felt more like a library than a home with everyone talking in hushed voices and being careful not to slam cupboards or clang dishes that might disturb Mom.

My brother and I ran loose and were becoming slightly feral, wearing the same jeans so many days in a row that the denim became more brown than blue, and bathing only when we remembered, which didn't seem to bother anyone because it was right and good to save water in drought-prone California. Which is why Matthew and I got in supremely big trouble when we got caught hiding behind the oak trees at the top of the driveway with the garden hose going full blast, dousing unsuspecting drivers with sudden rainstorms. It was bad enough we'd pulled a dangerous prank, but it was even worse that we wasted precious water with a looming drought. Grandpa was letting his fruit trees die, and he was worried that there wouldn't be enough flowers for his bees to make honey. Neighbors were rescuing gasping steelhead trout from what was left of the Carmel River, transferring them into water tanks in the back beds of their pickups and driving the fish to the mouth of the river, closer to the ocean, to release them.

I tried arguing that we had crimped the hose in between cars, but it didn't win any points. Granny ordered Grandpa to spank us anyway. But he did it in a way that was more symbolic than painful, making a big showy swing with his arm and slowing to a pat by the time his hand reached our bums. But we yowled from the shame of it all.

The real lesson we learned from the spanking was that our grandparents were exact opposites. She was the disciplinarian, and he was the softie. When they shared the newspaper in the morning, she fretted over the political news and he laughed at the comics. She worried about reputation and appearances; he wore tattered undershirts dribbled with coffee stains and never bothered to clean the black grime from under his fingernails. She was tidy; he never threw anything away, collecting his possessions into indoor and outdoor piles that grew taller and thicker by the year, which in a certain light matched the professional definition of hoarding. She detested the outdoors; he had to be coaxed inside.

When Granny met Grandpa during a square dance at the elementary school in Carmel Valley, she was a forty-year-old single mother living in the little red house with Mom, who was then nineteen. Barely a few months divorced, Granny was trying to socialize again, and Grandpa, three years younger, was perfectly satisfied being single. When Grandpa twirled Granny around, she noticed the strength in his upper body, the care he took to get the steps correct. It didn't hurt that she'd read about him already in Big Sur's monthly newsletter, *The Roundup*, which dubbed him Big Sur's Handsome Bachelor.

Grandpa wasn't looking for a mate; he was just fine with his bees, and he earned a steady income as a plumber, learning from friends how to make water flow to remote cabins where there was no centralized water system; digging wells and climbing the steep Santa Lucia Mountains to divert natural springs and creeks to homes below.

Ruth and Franklin were an odd couple but a good dancing pair, and began attending square dances together, even traveling to the faraway ones in Salinas and Sacramento.

On their third date, at a square dance in South Lake Tahoe, Granny asked him what his intentions were, and when he tried to dodge her question, she literally told him to "fish or cut bait." No one had ever confronted him so directly, and he was impressed. He agreed to marry her, and she convinced him right then and there to drive across the border into Nevada so they could tie the knot immediately, giving him no time to change his mind. They drove until they located a Carson City courthouse that offered around-the-clock weddings, summoned a janitor to serve as witness, and at nine that night became husband and wife. Mom was a little surprised and somewhat dubious of her sudden stepfather, but she didn't have time to get to know Grandpa. Four months after he moved in, she transferred from Monterey Peninsula College to study sociology at California State University, Fresno.

My grandparents knew scant little about one another when they married, but over time they learned to love their differences. He liked a cold beer; she preferred Manhattans. He spoke only when he had something to say; she spoke in monologues. But they fit, mainly because she liked to lead and he, averse to confrontation, willingly followed. He had no interest in power, prestige or money, and handed his income to Granny so she could figure out the bills and the taxes. They parted every morning for their separate worlds—hers in the classroom, his in the Big Sur wilderness—and then came together every night at the dinner table where he ate in silence as she lectured on a never-ending list of topics. Grandpa admired her mind, although he also had an Olympian appetite and could fill his plate four times in one sitting. This made him an excellent listener.

It didn't take long for Matthew and me to adjust to the rhythms of our grandparents' schedules. Granny preferred her afternoon cocktail lying down. After a full day of teaching grammar and arithmetic to a roomful of trying fifth-graders, her first order of business was to mix a Manhattan and recline on the orange shag rug in the living room, her head propped on a pillow and a newspaper spread before her. By now she had taught me how to make her drink, and I liked the daily ritual of it almost as much as she did. I poured brown bourbon into a tall blue plastic tumbler until it was two fingers high, splashed in some sweet vermouth from the green glass bottle and added two ice cubes and a neon red maraschino cherry. I swirled it around with a dinner spoon and brought it to her.

"Grazie," she said, reaching up from the floor.

With a loud licking of her fingers, she flipped the pages of the free *Carmel Pinecone* that she'd picked up at Jim's Market and told anyone within earshot what she thought about local politics.

"Goddammit all to hell, I can't believe they want to put streetlights in the village! Excuse my French."

Her outbursts were not invitations to respond. She kept her head down and continued her conversation of one.

"What do we need lights for? We don't even have any sidewalks. Damn Monterey County supervisors!" she said, taking another gulp from her tumbler. Outsider politicians were always trying to modernize unincorporated Carmel

Valley Village and ruin the reason people moved out to the country in the first place, she said.

I kept listening as I climbed into Grandpa's recliner and wiggled the handle on the side, trying to get the chair to go flat. I believed Granny was exceptionally smart, and knew things that regular people didn't. My opinion came from two sources: Granny herself, who had told me several times that her 140 score on an IQ test proved she was a genius; and secondly that she could predict the weather. I didn't know that forecasts were printed in the newspaper, so when I'd ask her what the weather was going to be like and she'd foresee sun or rain or frost, I thought she had some direct line to the universe.

She dropped phrases in Latin and Italian every once in a while, which sounded cosmopolitan to me. As the cocktail hours piled up, I was slowly starting to adopt her worldview, dividing people into those who were wrong and those who were right. I didn't know what a Democrat or a Republican was, but I had heard the words so often that I knew we were on the Democrat team. Granny's world was black and white, and therefore easy to follow. She was right, and anyone who disagreed was dim-witted and therefore deserved our pity.

"It's tedious being smart," she'd sigh, swirling the ice in her drink. "Waiting for everyone else to catch up to you. One day you'll know what I'm talking about."

Granny was now reading about the gasoline shortage and flipping the pages with more force. I went to the kitchen and helped myself to one of her cocktail cherries, and then slipped away to Mom's bedroom. The door, as usual, was shut, and there was no sound from inside. Mom had been in bed so long that she was becoming shimmery around the edges like a memory. I felt my mother more than I saw her, when she curled her body around me at night.

"Mom?"

I tapped lightly on the bedroom door. Nothing. I knocked a little harder. Her voice sounded like it came from under the covers, thick and muffled.

"Go away."

Her words pinched, and I winced reflexively. Mom still liked me; I knew that. I reminded myself that she just wasn't herself right now. Granny rounded the corner and spotted me lingering where I wasn't supposed to be. "Come with me," she said, placing a hand in the small of my back and guiding me to the kitchen. She lifted a wicker basket of wet clothes off the counter, and I followed her outdoors to hang the laundry. She dropped the basket on the ground with a thump under the wire clothesline that Grandpa had strung up between two T's made out of plumbing pipes.

"Hand me the clothes," she ordered. "I can't bend down on account of my bad back."

I passed her one of Grandpa's white cotton undershirts, encrusted with drips of plumber's putty and worn so thin I could see through the fabric. She snapped it into the wind once, then pinned it with clothespins. Then she reached toward me for the next item. I pulled out her floor-length quilted nightgown, the one covered in pink roses.

She cleared her throat.

"You know your mother is going to need everybody's help to get better," she said, contemplating the clothing in her hands. I knew what was coming. I was in trouble for knocking on the bedroom door again.

"I just needed Morris."

Granny paused and faced me.

"Aren't you getting a little old for a teddy bear?"

Her words were so horrible that I momentarily forgot what I was doing and dropped my favorite green-checkered dress on the ground. I couldn't sleep without Morris tucked in my arms. He was my only possession, the only thing left from Before.

"Dad gave him to me!"

Granny bent down to pick up my dress, and she grunted like it really hurt. It looked like she was stuck, but she put her hand to her back and rose slowly, puffing out her cheeks with the effort. She shook the dirt off my dress and continued pinning.

"That's another thing," she said. "I don't want you and Matthew mentioning your father around her. It only upsets her."

Dad was the only thing I wanted to talk about, but his name had not come up once since we landed in California. Everyone acted as if Dad didn't exist, and I was beginning to wonder if Matthew even remembered him. He had even started to refer to Grandpa as Daddy. Each time, Grandpa gently reminded him that he was a grandpa, not a daddy. It was like our life in Rhode Island was a movie, and the movie had ended, and that was that. Over and forgotten. If everyone pretends your dad doesn't exist, does he?

Granny was staring at me, waiting for me to agree to never say Dad's name. It was pointless to argue, because I would be taking Dad's side against hers and that would have repercussions I could only shudder to imagine. It's true I wanted Mom to get better. I didn't want to keep thinking of her as a sick person, someone with a weak heart and faraway eyes. I wanted her to braid my hair again, read *Winnie-the-Pooh* to me, take me with her to the grocery store. If that meant having silent conversations about Dad in my head, then that's what I would do. But before I submitted to Granny's ultimatum, I had to ask a question.

"When's he coming?"

Granny reached into her shirt pocket and pulled out a pack of cigarettes. She shook one out, lit it and relaxed her shoulders with the first exhale. She stared at the honey bus as if searching it for my answer.

"Your father is not a very good man," she said, keeping the back of her head to me. Then she indicated to me to hand her the next thing in the basket. Conversation over.

I put my tongue between my teeth to keep from calling Granny a liar. How dare she pick sides, as if she could just snip Dad out of my life with a swipe of her scissors? I had bat-ears; I knew that she talked about Dad with Mom sometimes, when their whispers floated out the gap at the bottom of the closed bedroom door. It wasn't right that they could talk about him but I couldn't—he was *my* father after all. I wasn't dumb; I'd figured out that Mom and Dad were having a fight and this wasn't a "visit" to California, but that didn't make my dad bad and my mom good. He was my dad, and he was coming back. Granny had everything all wrong.

The sun was low in the sky, and the honey bus looked

stage-lit with orange and yellow bulbs. Through the windows I could make out the shapes of three men crowded inside with Grandpa, passing honeycomb frames between them and shouting over the rat-a-tat of the machines inside.

I crept forward to get a closer look. The men had taken off their shirts in the heat and tied them to the overhead handrail. I couldn't hear what they were saying, but could tell they were swapping jokes, slapping one another on the back and doubling over with laughter. The men had an action-figure quality to them, their barrel chests rippling and shining with sweat as they hefted hive boxes and stacked jars of honey into towering pyramids. I studied their every move, even how their Adam's apples bobbed with each swig of beer, and I silently willed them to wave me inside with a swing of their Popeye arms. These were the Big Sur friends Grandpa grew up with, the ones who had taught him to rope cattle and dive with a snorkel for the iridescent abalone shells that I had found in the backyard. These were big men with big hands who showed Grandpa how to build log cabins from redwood trees, how to hunt wild boar, or clear landslides off the coast highway with heavy equipment. They were living Paul Bunyans, the Big Sur mountain men who fended for themselves in the wild.

I patted down the tall weeds and made a little burrow for myself where I could sit and watch them work. They used thick, heavy knives blackened with burnt sugars to gently slice open wax-sealed honeycomb, exposing the orange honey underneath. They lowered honeycomb frames into the massive spinner, and cranked a handle protruding above it from left to right, using two hands and all their body weight to shift its position. I saw one of the men yank on a pull-cord several times and heard the lawn-mower motor sputter to life. The flywheel started to rotate and whine, and as it picked up speed, the bus began to rock slightly from side to side. The pump kicked in and forced the honey from the bottom of the extractor, up through the overhead pipes, and directed it to cascade in two streams into the holding tanks. It was nothing short of miraculous; like striking gold.

I stayed in my spot until the sun slipped behind the ridgeline and the crickets came out to sing. The men flicked on the construction lights in the bus and hung them from the handrails so they could keep working into the night.

I was drawn to the bus like a moth to flame, by an irrepressible longing that I felt as a physical ache, a gnawing in my belly to disappear into the secluded protection of an enclosed space like a submarine, or a bus. The honey bus looked like it was warm inside, and safe. I wanted the men to invite me to join their secret club, and to teach me how to make something beautiful with my hands. My pulse sped up when I watched them work together in a harmony of familiar dance movements, passing frames of dripping honeycomb between them and taking turns capturing the honey into glass jars as it flowed out of the spouts. I could tell the bus made them happy, and I believed it could do the same for me.

I was struck by a certainty, from some deep place inside myself, that something important was waiting for me in the bus, like the answer to a question that I hadn't yet asked.

All I had to do was get inside.

3

The Secret Language of Bees

1975—Late Spring

I didn't limit my snooping to the outdoors. I brazenly opened drawers, rifled through closets, and took a keen interest in what Granny and Grandpa had tucked away inside the house. Because my grandparents were old people, their stuff was old, too, and I enjoyed hunting for rare artifacts forgotten in the far corners of their history. I found arrowheads that Grandpa had unearthed while digging pipelines in Big Sur, and inside the cedar chest I dusted off a stack of *LIFE* magazines with JFK, Elvis, and the first astronauts on the covers. The kitchen cupboards held a boneyard of cooking gadgets that Granny had tried once and then deemed ridiculous.

One morning I dug out an Osterizer blender from deep in the back of the cabinet under the sink. I wedged the glass pitcher onto the base, put the lid on, pressed one of the buttons and it whined to life. For a bored girl with few toys,

I suddenly possessed this most miraculous machine and a whole kitchen packed with mystifying things pickled in mason jars. I opened the pantry and selected a jar containing a bright green Jell-O-looking substance, unscrewed the lid and sniffed: mint jelly. That could taste good—I liked mint gum, as well as jelly on toast—so I scooped it into the blender and added milk. Figuring I needed more than two things to make a smoothie, I did another quick scan of the kitchen until my eyes rested on the cereal boxes lined up on top of the fridge. I dragged the stool over and pulled down the corn flakes, thinking it would make my drink thicker. I pressed the button for the highest speed and whirred it into a concoction resembling runny, lumpy toothpaste, which I poured into a ceramic mug and brought to Grandpa, who was at the dining room table watching the birds peck at seed he'd sprinkled on the deck railing.

Grandpa would eat anything. He chewed chicken gizzards, said cow tongue was so delicious it put hair on his chest, and devoured artichoke leaves whole. He'd even developed a technique to pull every kernel clean off an ear of corn, using only his lower teeth and running the cob back and forth before his mouth like the carriage return on a typewriter. I presented him with my milkshake. He took a swig and then needed a few seconds to come up with an adjective.

"Refreshing!" he said, chasing it down with coffee. "What's it called?"

"Mintshake," I said.

He nodded thoughtfully and strummed his fingers on the table, like a gourmand considering a tasting note.

"Let's share it," he said, sliding the cup back toward me.

It was a dare, all right. I could tell Grandpa was trying to keep a straight face as I reached for it, but just as I was about to take a drink, a low hum distracted us from our standoff. Grandpa reflexively turned toward the sound and tracked something flying in the air. I followed his gaze until I saw what he did—a honeybee hovering over the dining room table. It was suspended in the air with its legs dangling beneath its body, keeping itself in place by beating its wings so fast they became invisible. I set the cup down and leaned back in slow motion. The bee, watching my every move, began to slowly come toward me, flying in slow arcs left and right, inching closer with each swing.

My muscles tensed, and I willed the bee to please, please, go take a hike. But it was attracted to the sugary smell inside my cup, and determined to have a taste. When it was about to land on the rim, I swatted at it.

The bee emitted a shrill *zzztttt!* in response, and zoomed in an anxious circle above our heads.

Grandpa jumped out of his chair and grabbed my forearm so tightly I could feel him pressing bone. I startled, frightened by the sudden aggressiveness of his touch. He'd never gotten mad at me before; he always fake-spanked Matthew and me when Granny forced him to punish us for misbehaving. He leaned toward me until we were nearly touching noses and locked eyes. His words were deliberate and forceful, each one like the clap of a church bell.

"You. Must. Never. Hurt. Bees." He didn't look away until he was certain his words had landed in my brain. I must have done something truly awful for Grandpa to scold me, but I was confused. Bees stung people. They were pests, like mosquitos. Who cares if I smashed one? Wouldn't I be doing the right thing by protecting myself?

"It was going to sting me!" I protested.

Grandpa's eyebrows sprang up in disbelief. "Why do you say that?"

The bee was now slamming itself into the window trying to fly away. Its buzz rose to a shriek. I thought perhaps we should be having this conversation in a different room, but Grandpa was unperturbed by the sight of a stinging insect going berserk. I kept one eye on the frenzied bee as I tried to answer Grandpa's question.

"Because bees always sting."

"Come here," Grandpa said.

I followed him into the kitchen, where he searched the cupboards until he found an empty honey jar.

"Go get a piece of paper," he said.

I was eager to do anything to get back on his good side. I raced to Granny's desk and pulled out a piece of her fancy stationery, and practically bowed as I offered it to him.

"Listen," he said, cupping his ear and cocking his head toward the buzz. "It's high-pitched," Grandpa said. "It's in distress. Do you see it?"

I followed the sound until I saw the bee gliding in a wobbly circle around the room, looking for a way out, until it rested on the dining room window facing the deck.

"There!" I pointed.

Grandpa crept softly toward it, hiding the jar behind his back. When he was directly behind the bee, he reached up and imprisoned it in one swift motion. With his free hand, he slipped the paper between the window and the mouth of the jar, forming a temporary lid. He stepped away, holding the trap in his hands, and the bee crawled up the glass, tapping the inside of the jar with its antennae.

"Okay, come get the door for me," he said.

We stepped outside together, and instead of releasing the bee, Grandpa sat on the back step and patted the space next to him, signaling me to sit near.

"Hold out your arm."

He tilted the jar as if he was going to release the bee onto my forearm. I jerked my hand back.

"It's going to sting me!" I wailed.

He sighed like he was summoning all his patience, and then turned to me again.

"Bees won't hurt you if you don't hurt them."

Most of my information about bees came from cartoons in which bees always traveled in bloodthirsty swarms terrorizing all manner of people, coyotes, pigs and rabbits. I mentioned this to Grandpa.

"That's make-believe," he said. "Honeybees don't go on the attack. They will only sting to defend their home. They know that if they sting they will die, so they'll give you plenty of warnings first."

Grandpa reached for my arm again, but I tucked it behind my back, still uncertain. The bee was now incensed, banging into the walls of its glass prison. Grandpa set the jar down and spoke to me slowly and carefully.

"Bees can talk, but not with words. You need to watch how they behave to understand their language. For example," he said, lifting a finger to numerate his points. "If you open a hive and hear a soft chewing sound, that means the bees are busy and happy. If you hear a roar, that means they are upset about something."

I watched the bee get more frantic by the second.

"Two," he said, holding up a second finger. "Bees will ask you to back away from the hive by head-butting you. It's a polite warning to step away so they don't have to sting you."

I was starting to understand that Grandpa might know bees in a different way than everybody else. He spent every day with them, so he probably could tell what they were thinking. But that didn't mean that I wanted a bee to crawl on me. I trusted Grandpa wouldn't do anything to hurt me, but I couldn't say the same for the trapped bee, who by the looks of things was now totally, royally, pissed. He reached for the jar again and brought it over to me. I shook my head no.

"You mustn't be afraid around bees," he said. "They can sense fear, and it will make them scared, too. But if you are calm, they will stay calm."

"I'm still scared," I whispered.

"The bee is more frightened of you," he said. "Can you imagine how scary it is to be this small in a world that is so big?"

He was right, I wouldn't want to change places with a bee. A little bit of my trepidation melted knowing the bee was also scared. I knew I wouldn't hurt it, but the bee couldn't know that for sure. I stretched my arm out again, ever so gently.

"You ready?"

I nodded as I watched the bee fall onto its back inside the jar, its six legs scrabbling to find footing.

"Bees are sensitive, so no sudden movements, and no

loud noises, okay? You must always move slowly and quietly around bees to make them feel safe."

I promised to hold still, an easy pact because I was too terrified to move. I tried to summon calming thoughts, but it was impossible to do on command. Grandpa tapped the jar on the underside of my wrist, and the bee tumbled out. It stood still as I held my breath, then it took a few tentative steps.

"Tickles," I whispered. This close, I could see that a honeybee's body was a miracle of miniature interlocking parts, like the insides of a watch. Its antennae, two L-shaped sticks that swiveled in sockets on its forehead between its eyes, searched the air and tapped on my skin, reminding me of a person without sight using a cane to get a mental picture of a place.

"What's it doing?"

"Checking you out," Grandpa said. "A bee's antennae can smell, feel and taste."

Imagine that. Having a body part that is a nose, fingertip and tongue together. As the bee got used to me, I got used to it. Grandpa was right. This small insect was not my enemy. I carefully lifted my arm until I could see into its eyes, shaped like two glossy black commas on the side of its head. Fear gave way to fascination as I studied how it was put together, so small, so perfect.

Veins crisscrossed its shimmering wings. It was furry, and its abdomen expanded and contracted with each breath. I looked closer at the stripes, and noticed that the orange bands had small hairs and the black ones were slick. The bee's legs tapered to tiny hooks, and it was now using its front two pair to stroke its antennae. Cleaning or scratching them, I guessed.

"What do you think?" Grandpa asked.

"Can I keep it?"

"'Fraid not. It will die of loneliness if you separate it from its hive."

I was beginning to understand that bees have emotions, like people, and like people they live in families where they feel safe and loved. They will lose their spirit if they don't have the security of their hive mates. I was about to ask if we should return this bee to its hive when it parted its mandibles and unfurled a long red tongue.

"It's going to bite me!" I shrieked.

"Shhhh, hold still," Grandpa whispered. The bee tasted my arm tentatively, realized that I was not a flower and recoiled its tongue. The bee put its hind end in the air and fanned its wings so rapidly that I could feel a vibration on my skin. Then it lifted off and was gone.

Grandpa stood, reached for my hand and pulled me to my feet.

"Meredith, never kill something unless you are going to eat it."

I gave him my word.

That night when I got under the sheets, Mom was already snoring. I cleared my throat hoping that would wake her, and when that didn't work, I jiggled the bed, just a little bit.

"Hmmmm?"

"Hey, Mom."

She grunted and turned toward me with eyes closed. "What?"

"Did you know bees die after they sting?"

"Shhhh. You'll wake your brother."

I lowered my voice and whispered.

"Their guts come out with the stinger."

"That's nice."

Mom rolled me away from her, then tucked her knees under mine and drew me to her stomach. I was about to brag about picking up a bee with my bare hands, but I felt her legs twitch and realized that she had fallen back asleep.

I lay there, my mind swimming with new questions about bees. Grandpa had just cracked open a portal to a secret microcosmos in our backyard, and now that I knew bees lived in families, I wanted to know everything about them. Which bees are the parents? How many bees in one family? How do they remember which hive they live in? What does it look like inside a beehive? Do they sleep at night? How do they make honey in there?

Grandpa had proven to me that I could get close to a honeybee without getting stung. I was coming around to the opinion that fearsome animals and insects rarely live up to the reputations foisted on them by circuses and monster movies. Grandpa was teaching Matthew and me that all creatures were sacred, with their own inner emotional lives. As part of our education, after dinner each night we climbed into the recliner with Grandpa to watch his favorite nature shows. I'd been astonished to watch male lions play with their cubs, aquarium octopuses reach from the water to embrace their human handlers, or elephants dig stairs leading out of a deep mudhole so a drowning baby could clamber to safety. So it made me wonder, what if bees were compassionate like that, and what if I could teach myself how to see it? As a girl needing to know that love existed naturally all around her, it was thrilling to realize that I didn't have to wait for Wild Kingdom or Jacques Cousteau to be reassured. The mysteries of the animal kingdom were within my reach, anytime I wanted. That night when I went to bed, the confines of our small room expanded ever so slightly. I had found one good thing—a reason how California might make me happy.

I awoke to the percolator bubbling on top of the stove, so I knew my grandparents were up. I tiptoed down the hall and pushed open their bedroom door. Granny was reading aloud to Grandpa from the *Monterey Herald* while he looked at the photos in a beekeeping magazine called *Gleanings in Bee Culture*. On weekends, they liked to ease into the day. I climbed onto their small four-poster bed, wedged myself in between them and asked Grandpa if he could show me his beehives.

"Whoa, Nelly," Grandpa said, putting down his magazine. "I haven't had my think-juice yet."

"Excellent point," Granny said. "Sounds like the coffee's done, Franklin."

Grandpa dutifully threw back the covers and slid his feet into slippers, and I heard his joints crack as he pushed himself upright. I sighed dramatically, but nobody acknowledged it. I was in for a long wait. On Saturdays and Sundays they savored several cups of coffee in bed, as Granny curated the newspaper front to back, reading aloud particularly important paragraphs to Grandpa, enhanced by her commentary. Grandpa would often get weary at a certain point, but he never complained. Instead, he would distract her by gripping sections of the paper with his strong toes and dropping the pages on her lap. Granny thought it was repulsive; Grandpa thought it was a riot.

I wandered outside and spotted Matthew lifting his chubby leg and stomping on something near the vegetable garden. When I got closer, I could see that he was killing snails. He smiled when he saw me approach, and lifted his shoe to display the slimy puddle he'd made on the ground. He was helping Grandpa, who'd shown him how to hunt the marauders who ate his crops. Snails and gophers were the only exceptions to Grandpa's no-kill rule.

"Gross," I said, slightly unnerved by how much my brother was enjoying himself.

He held a snail between a thumb and forefinger and dropped it on the ground.

"You do it," he commanded.

I reached for his hand instead. "Come on, I have another job for you."

His eyes widened, and he bounced alongside me as I walked him toward the honey bus. There was about a foot and a half of clearance under the chassis. If we crawled underneath, we could hopefully find a rusted-out hole or some type of entry and maybe climb through to get inside the bus. I'd already tried pushing all the windows, and had inserted all manner of sticks and screwdrivers and butter knives inside the opening where the back door handle used to be, hoping to pop the lock. This was my last idea. I figured I'd need Matthew if we found an opening too small for me.

I slid under first on my back, as it was more Matthew's nature to see if something was safe before trying it. He watched my legs disappear and waited for my report. A tangle of weeds blocked my view of the undercarriage, so I used a snow angel technique to knock them down. I pressed here and there on the bus floor with my foot to test for weak spots. The metal was rusty, but solid. I kicked at the exhaust pipe, and it rattled some, showering me in fine dirt. I scooted toward the front of the bus, and bumped into a discarded tire. Other than that, the only thing I found under the bus was a graveyard of corroded five-gallon Wesson Oil cans.

I gave up searching and rested for a moment on my back, trying to think. There had to be a solution that I was overlooking. Matthew called out to me, and when I turned to look over my shoulder, I saw him on his hands and knees peering under the bus. Then two legs appeared and framed my brother.

"What's so interesting under there?" I heard Grandpa ask my brother.

"Mare-miss," my brother said, pointing. His tongue still hadn't mastered my three-syllable name.

Grandpa got down on his belly next to Matthew, and now both of them were staring. I held still because I felt like I had just been caught doing something, not anything bad, just something slightly embarrassing.

"Whatcha doing under there?"

"Trying to get in."

"Don't you know the door's up here?"

"It's locked."

"To keep little kids out."

Grandpa reached under the bus and crooked his finger, signaling me to come to him. I scrabbled out, and as he helped me to my feet, he brushed the dirt from my back and plucked off the burrs. Whatever was in the bus would have to wait. Until I got bigger, whenever that was. The only people admitted entry were Grandpa's friends, so I imagined I would have to wait until I was an adult, which might as well be never.

"I thought you wanted to see the bees," Grandpa said.

His counteroffer was exquisitely played, and I perked up immediately. As my part of the deal, I had to come inside for breakfast first.

Belly properly filled with pancakes, I followed Grandpa to the back fence, where he kept a row of six beehives. The sun was shining on the slit entrances at the base of the hives, illuminating the landing boards where the bees were flitting in and out. A small cloud of bees hovered before each hive, all the foragers waiting for a clear shot to get back inside. I noticed that the bees were buzzing in a different way than the one we caught in the house; their sound didn't have the urgency of a shout, it was more contented and calm like a person humming a song. I stood in front of the right-most hive, about a foot away from the entrance so I could watch them. I felt Grandpa's hand on my shoulder.

"Don't stand there," he said. "See what's happening behind you?"

I turned and saw a traffic jam of bees jiggling in the air, unwilling to go around me to get into the hive. The backup was growing by the second.

"You're in their flight path," he said, guiding me to the side of the hive. As soon as I stepped out of the way, the clot of waiting bees whooshed in a comet back to their hive. I knelt down next to the hive so I was eye level with the bees.

One by one they marched to the entrance, cleaned their antennae, then crouched and launched like a jet fighter.

"What do you see?"

"Lots of bees coming and going," I said.

"Look closer."

I did, and saw the same thing. Bees flying in. Bees flying out. So many it was hard to keep my eye on one bee at a time. Grandpa took a comb out of his back pocket and whisked it through his hair in three practiced swipes, top and sides, waiting for me to see whatever it was I was supposed to see. Then he pointed at the landing board. "Yellow!" he announced.

All I saw were bees.

"There's orange! Gray! Yellow again!"

And then I saw it. Some of the returning bees had something stuck to their back legs. Every fifth or sixth honeybee that returned waddled in carrying small balls like the pills that collect on a favorite sweater, some loads no bigger than the head of a pin, others the size of a lentil, so large the bee strained under the weight.

"What is it?"

"Pollen. From flowers. The color tells you which flower they came from. Tan is from the almond tree. Gray is the blackberries. Orange is poppy. Yellow is mustard, most likely."

"What's it for?"

"Bee bread."

Now he was just messing with me. Bees can't bake bread. All they make is honey. Everybody knew that.

"Grandpa!"

"What? You don't believe me?"

"No."

"Suit yourself. Bees mix the pollen with a little spit and nectar and feed it to their babies. Bee bread."

It made some sense, but it was just too weird. I waited for him to giggle at his own joke, but he kept a straight face. Grandpa had told the truth when he said it was safe to let a bee crawl on me, so I guessed he knew what pollen was for. For the moment, I played along.

"They're making bread in there?"

"They push the pollen off their legs, chew it with nectar and store it in the honeycomb."

"Can I see?"

"Not today. I don't want to disturb them right now. They are building new wax."

Just then the fattest bee I'd ever seen lumbered out of the hive. It was wider and stockier than all the others, and its head was comprised almost entirely of two massive eyes. I watched it approach several of the regular-size bees and tap its antennae against theirs. Every bee it touched backed up and walked around it, as if irritated by being bumped.

"Is that the queen bee?"

Grandpa picked it up and put it in his palm. "Nope. A drone…a boy bee. He's begging for food."

I asked Grandpa why he didn't just get his own food.

"Boy bees don't do any work. All those bees you see with pollen? All girls. Boys don't collect nectar or pollen for the hive, they don't feed the babies and they don't make wax or honey. They don't even have stingers, so they can't protect the hive."

Grandpa returned the drone to the hive entrance where it

resumed the search for handouts. Finally, one of the returning girl bees paused and linked tongues with him. Feeding him nectar, Grandpa said.

"He only has one job, but I'll explain it to you when you're a little older."

Grandpa had set up two stumps near his apiary, and we took seats and watched the bees flying as one would watch a fire, or the sea, lulled by all the individual movements that combined together into a single flow. I liked interpreting the patterns of their routine, to know that the bees weren't just flying willy-nilly; there was an order to what they were doing. They were out grocery shopping for bread and nectar. A beehive could seem chaotic if you didn't understand that bees had a plan for everything.

I could have never guessed that a beehive is a female place, a castle with a queen but no king. All the worker bees inside are female; around sixty thousand daughters that look after their mother by feeding her, bringing her water droplets and keeping her warm at night. The colony would wither and die without a queen laying eggs. Yet without her daughters taking care of her, the queen would either starve or freeze to death.

Their need for one another was what kept them strong.

4

Homecoming

1975—Summer

Our grandparents had the incredible good fortune to live only steps away from the Carmel Valley Airfield, where two-seater planes landed and took off a handful of times a month. It was nothing more than a dirt landing strip, with just one runway and a taxiway, and no lights, fences or security of any kind. There were no markings or signs to direct pilots and the tattered wind sock was rendered useless. Pilots had to radio in to a neighbor with a view of the runway and ask which way the wind was blowing.

Uprooted as we were without access to our playthings or our former playmates, Matthew and I had to get creative in our diversions and make use of whatever was readily available. We tried to build pyramids with Granny's poker cards, we put out birdseed and waited for birds, but an airport with real live planes was an entertainment jackpot.

All it took was the rumble of an incoming propeller, and Matthew would drop whatever was in his hand and go streaking out of the house to look for a plane. He was mad for those planes, falling into a near trance as he watched them come in for a landing. He'd run for Grandpa and yank him by the hand, urging him to take us across the street so we could stand by the runway and feel the wind wash over us as the plane whooshed down from the skies.

One afternoon we heard the telltale engine noise, but Grandpa was working in Big Sur and we didn't have an escort. But now that we were spending so much time alone together, a budding solidarity was forming between Matthew and me, and sometimes our companionship crossed over into mischief. We hesitated ever so slightly, looked back to the quiet house, then grinned at one another and bolted across the road, huffing it up the small incline to reach the airstrip just as the plane was circling overhead.

Matthew wanted to get closer to the plane this time, so we crept to the median between the two runways and sat down in the grass to wait for the plane to fly over us. I snapped off a mustard blossom and ate it, like I'd seen Grandpa do. I offered a yellow bloom to Matthew, but he wrinkled his nose. We could hear the propeller approaching, beating the air like thunder. Matthew reached for my hand, and we stretched out on our backs and looked skyward.

When the plane's underbelly crossed not twenty feet overhead, we felt the growling engine in our chests and screamed with the same mix of joy and terror that roller coasters were designed for. I can't imagine what the pilot must have thought when he saw two small children pop into view at the last minute. We waved, innocently hoping he'd see us; he probably had heart palpitations.

We sat up and watched the plane make a few short screeching hops and then touch down. It rolled toward the end of the landing strip where a collection of similar planes was parked, their wings chained to the ground.

Just then the plane, its blades still whirring, made a U-turn and started slowly approaching us. It was halfway down the runway when the plane stopped and the pilot got out and shouted something at us. We couldn't hear the words, but picked out the unmistakable tone of an adult who "would like a word" with us. We sprang to our feet and took off, and before I could count to ten, we were back behind our little red house, bent over, sucking in oxygen. I hoped the pilot hadn't seen which house we ran to, and secretly promised myself never to do that again.

When we caught our breath, we walked as innocently as possible into the kitchen, where Granny was scorching something in the electric skillet. She'd given up on the oven long ago, insisting the temperature dial had a manufacturer's defect that burned her food. The oven became a tabletop for a square electric frying pan no bigger than a pizza box, and although her cursing had subsided considerably, every breakfast, lunch and dinner still came out blackened and overdone.

"Where have you two been?" she asked, keeping her back to us and furiously scraping at something with the spatula. I put my finger to my lips to remind Matthew we couldn't tell. He nodded.

"Nowhere. Just outside," I said.

"Well, stay close. Dinner's almost ready."

"We saw a plane!" Matthew piped. The kid just couldn't help himself. Before the conversation could progress, I quickly grabbed his hand and led him to the living room, distracting him with a suggestion that we build a fort.

Granny had one of those couches that felt as long as a Cadillac, made with two rectangular bottom cushions that when removed made excellent walls. We dismantled the stuffed yellow chair for the roof pieces and assembled a hut in front of the television, leaving a peephole so we could sit inside and watch TV. It was almost like being in the dark of a real movie theater. We settled in to watch Matthew's favorite show, *Emergency!*, about two Los Angeles paramedics who carry a hospital phone in a box and rescue accident victims, mostly by jolting them back to life with electrical paddles.

"TV's too loud!" Granny called from the kitchen.

Just then a car exploded on-screen at full volume.

I was cozy. I didn't feel like removing a wall and crawling *all the way* to the TV to reach the volume knob.

"Turn the TV down," I said to Matthew.

He ignored me. Lately, Matthew's adoration of me seemed to be waning. This was disturbing on two counts. One, he was no longer following my orders. The other day he'd even refused to let me put every necklace and bangle in Mom's jewelry box on him, something we did all the time. But worse, he was all I had left of my family, and I couldn't tolerate the thought of him leaving me, too. I tried not to take his emerging independence personally, it was part of his growing up after all, but I was afraid it signaled something deeper, that he one day wouldn't need me. The thought of Matthew leaving me was so terrifying that I became meaner to try to keep him in line, to show him that there were severe consequences for disobeying me. So if he wasn't going to turn the TV down, then he wasn't going to get to stay in the hut, either. I knocked the sofa cushion nearest me, and our house toppled on us. Matthew howled in outrage as he kicked his way free of the ruins.

Granny appeared in the living room, wiping her hands on a dish towel. She shot us a look that said we were riding her last nerve. Then she cranked down the volume, and that's when we heard someone knocking on the front door.

How long the visitor had been trying to get our attention, we couldn't say. Most likely it was one of Grandpa's honey customers, dropping by unannounced with an empty glass jar in hand. Grandpa wasn't home, so whoever it was would have to leave their jar on the doorstep with a check or cash in it, and Grandpa would swap the money with honey, then put it back outside so they could fetch it later.

Granny opened the door, and I saw her back stiffen.

Then she shouted my mother's name over her shoulder. *"Sal-leeeee!"*

I heard the creak of the bedroom door, and Mom padded into the living room in rumpled sweatpants and T-shirt, an outfit that doubled as her nightgown.

"You don't have to yell, Mom," she said, blinking in the afternoon light. Mom came up behind Granny and put one arm on the door frame and leaned in. Then she took a step back.

"David," she said.

I heard a low male voice, and the hair on the back of my neck prickled up.

Dad!

The vault inside me where I stored all my secret thoughts about Dad flung open, and fireworks exploded out of every pore. Six months of wishing in the lonely quiet of night had worked its magic, and now everything was going to snap back to the way it was before, just like I knew it always would.

I clicked off the TV, and Dad's silky words swirled into the living room, wrapping me in a tight fabric and pulling me toward him. I knew he would come back. Now we could finally go home, Mom would be happy again, and Matthew and I would get our own rooms back. I looked at my little brother, and he was bouncing up and down, his eyes fixed on the door.

"Daddy, Daddy, Daddy, Daddy!" he sang.

I leaped in the direction of Dad's voice, but Mom and Granny wouldn't step aside or open the door wider than a couple of inches, so all I could see were bits of him—the side of his leather Top-Sider shoe, a patch of ink-black hair. I peered through the crack in the door and spotted our green Volvo parked in the driveway by the eucalyptus tree. *He must really want us back if he drove all the way here*, I thought.

"Did you bring my portable dishwasher?" Mom demanded. "The kids' toys?"

I tugged on Granny's sleeve, but she didn't respond. I tapped on Mom's back. Nothing.

My father had just driven across country in Mom's Volvo to return it to her, and no one had explained this beforehand to Matthew and me. He'd stayed the night at his mother's house in Pacific Grove, and he'd asked her to follow him to our house the next day and park a few streets away so he could catch a ride back to the airport. He'd anticipated a possible confrontation at our house and wanted to spare his mother from seeing it, so they made a plan for him to walk to the village where there was a one-block strip with a grocery, barbershop, bank and restaurant, and meet her in the parking lot.

I knew none of this. When Dad suddenly appeared on our doorstep, I'd assumed that he was there to fetch us. I looked on, stunned, as Granny blocked him from coming through the door.

Something wasn't right. Dad must know we were in the house, so why wasn't he coming inside? What was taking so long? Why weren't they letting him in? Granny was speaking in clipped sentences, with the same undertone of disgust she saved for bad politicians she read about in the newspapers. I heard Dad mumbling, like he was apologizing, and the air became thick with malice. Their voices were getting louder, darker, sharper, and my muscles clenched with the memory of our last night in Rhode Island. Then Mom's voiced cracked into thunder.

"How can you do this to me?" Mom shrieked. "Don't you care about your own children?"

Dad's fingers flashed into the house and dropped the car keys into Granny's open palm. She tossed the key ring a few feet onto her writing desk as if it was a stinky shoe that she didn't want to touch. Mom stepped outside to talk to Dad and Granny closed the door, thumping her butt into it to make sure the latch caught. She pushed the button in the doorknob to lock it, and swiped her palms together in a gesture of finishing something, of wiping flour off her hands. She walked back to the kitchen without glancing our way, as if nothing had happened.

Things were moving too fast. I could hear Mom outside roaring at Dad. I didn't know what *divorce* meant, but I caught the finality in her voice as she spat the word at him, and that told me all I needed to know—that whatever was wrong with my parents was unfixable.

"Don't you want your kids?" she wailed.

Matthew looked at me with wide eyes, searching my face for reassurance. I took a step closer to him, and he wrapped his arm around my leg.

I heard Dad's voice rise to meet Mom's, and they became two dogs, barking and growling at each other. A familiar dread pressed down on my rib cage, and I understood that if I didn't get through that door, I might never see my father again. This was my one chance to try to change his mind. Maybe, if he saw me, if I pleaded with him, he'd stay. I couldn't let Dad come this close and then slip away without trying. I lunged for the door, unlocking it just in time to see Dad walking down the driveway and toward the road. The neighborhood reverberated with Mom's voice as she hollered to his back.

"Mark my words, YOU WILL REGRET THIS!"

I opened my mouth to scream, but cobwebs caught my voice. I tried to run, but my legs were somehow wrapped in iron chains. Mom picked up Matthew and jogged after Dad, chastising him for leaving his family.

My brain and my body had somehow disconnected, and I could no longer discern what was real and what was only in my imagination. Dad continued walking with his eyes fixed forward. When he was nearly to the road, the blood rushed back into my legs and I raced to the top of the driveway, where Mom was standing with Matthew on her hip, watching Dad go. She had quieted now and stood mute, as if she, too, couldn't grasp what she was witnessing.

My mind spun frantically, searching for an explanation. Then it suddenly rested on a simple solution, and a butterfly of hope flittered down and landed on my shoulder. This was all a bad dream. Since moving to California, I'd started having nightmares, so I tried to convince myself that I was going to wake up from this.

Dad was getting smaller with each step. I started walking after him, and Mom reached down and held me back. Her fingertips pressed on my chest, and I could feel the message inside them: there's nothing you can do. My pulse quickened as I realized I had run out of time. This was real, and Dad was leaving for good.

Hot tears welled in my eyes, and Dad became a blurry smudge. I wept in a way I never knew possible, the sobs chuffing out of me in painful bursts. My tears fell to the pavement, leaving little dark circles, and Matthew swiveled around in Mom's arms to see what was the matter with me. He probably wouldn't remember this day, and that made me cry even harder.

Dad heard me. He turned around and began walking back. I stopped breathing, waiting. When he reached us, he fell on one knee and hugged me so hard that I coughed for air. I detected the sweet, raisin-aroma of his perspiration and he was trembling, like his whole body was crying. I scanned him as if I'd never seen him before, rememorizing the dark hair covering his forearms, the stretchy gold band of his watch. There was a tan line where his wedding ring used to be.

"I will always be your dad," he whispered into my ear. I let myself melt onto his chest so I couldn't feel my edges anymore. I wanted to tell him to stay, but there was no room for words in between my sobs. I couldn't control anything anymore, not even language.

"I love you," he said, squeezing me once more and releasing me. He stood and took one last look at Matthew and Mom, and began walking down Via Contenta again. Mom tugged on my arm.

"Let's go."

I yanked my hand out of hers and started walking after Dad. I made it as far as the neighbor's house when I realized I was powerless to stop him from shrinking into the distance.

Mom left me then, and fled back to the house with my brother bouncing in her arms.

I stayed on the road and watched Dad reach the corner, turn left and blink out of sight. My vision narrowed, and I forced all my energy on the spot where Dad had been just seconds ago, as if wishing could bring him back. I wished so hard that I felt light-headed, like I was going to faint.

The completeness of my fate pressed down on me, and I stumbled back home, my body so numb I couldn't feel the ground beneath me. I needed Mom. I was desperate to curl inside the curve of her and have her tell me it was only a bad dream. I wanted her to tell me Dad was going to the store, and everything was going to be all right. There had to be more time, a second chance. I ran through the house, looking for her, and finally stopped before the closed bedroom door.

I knocked.

"Mom?"

She didn't make a sound, so I slowly turned the doorknob and wedged the door a crack. A curl of cigarette smoke wafted out.

"Mom?"

I heard, but could not see, her shift in the sheets.

"Not now, Meredith."

Her pale fingers reached out of the dark and tapped her cigarette into the ashtray on the headboard. I knew I had been dismissed, but my legs stayed rooted to the threshold. She exhaled, then swept the sheets off with one arm and swung herself into a sitting position. She came toward me, a moving shadow in the smoky haze. I lifted my arms expectantly.

She shut the door.

I felt my knees buckle and stumbled, catching myself against the wall.

"Meredith! I hope that's not you I hear bothering your mother!" Granny called out over the sizzling skillet.

The reptilian part of my brain sent one instruction: flee. I wanted to disappear, to get away from everybody and everything—to crawl into a dark hole and scream. I pushed myself off the wall, streaked out the kitchen door and bolted outside.

I heard the slender leaves of the eucalyptus tree shushing in the breeze. The massive tree stood taller than our house, and its summer bloom had appeared seemingly overnight. Grandpa's honeybees were losing their minds inside its butter-scented blossoms, scrabbling and rolling in the yellow pollen. Tens of thousands of bees buzzed in such a chorus that it sounded like all the overhead power lines were sizzling.

I felt an uncontrollable urge to get closer to the bees.

My legs, without consulting the rest of me, began walking toward the tree. I placed my hand on the curling outer skin of the trunk and felt a faint pulse, like sound waves coming from a stereo speaker. Then, as if someone else had taken control of my muscles, I watched my right sneaker wedge itself into the deep groove of the double trunk, and I hoisted myself up limb by limb, climbing higher and higher into the hum until I was completely concealed in a cloud of honeybees.

I reclined into a crook of an uppermost branch and watched the bees dart before me like sideways rain, so intent on their free buffet that they hardly noticed a girl in their midst. This close, I could see the blossoms were shaped like miniature hula skirts with a hard shell on top and a ring of delicate fronds. The bees swam in the center of them, wriggling their legs in a frenzied crawl stroke to coat themselves with yellow dust.

Bees circled me, their song stronger now. I stayed very still, letting the bees become accustomed to my presence.

When one landed on my leg, I simply watched it, holding my breath until it flew away. When it happened a second and third time, I began to trust that the bee was only resting and wouldn't harm me.

I studied the bees as they pushed pollen grains onto their back pair of legs, packing the granules down into tight, round saddlebags. I noticed they used their forelegs to brush pollen dust off their eyes and antennae, working from front to back, first cleaning their triangular head, then pushing the dust down their bodies toward their abdomens, then finally shoving the grains onto their back set of legs, packing the yellow grains down into two concave pockets designed to hold pollen. The bees took their time, and when their pollen cargo felt just right, they zoomed back to the hive to store it in their honeycomb pantry.

I inhaled the menthol of the eucalyptus and felt my outlines melting away. I was safe inside a buzzing force field where no one could see me and no one had to feel sorry for me. Above ground, I was no longer that girl without a father at home. I wasn't the girl whose mother never got out of bed. The bees made me invisible. I closed my eyes and let myself be lulled by their hymn.

The sun went down, the bees went home for the night, but still I stayed in the tree. I didn't want to come back down to the ground. Down there was chaos. Up here, the bees turned chaos into order. Up here, there was an entire species living its own life, oblivious to the fog of depression that engulfed our house. The bees reminded me that the world was so much larger than my family's insular problems. I liked being this close to creatures that were relentless about their work, natural survivors that avoided self-pity and never gave up.

I felt a compulsion to be near bees that I couldn't explain. On a deep level, the bees were teaching me the importance of taking care of myself. I could see, with my very own eyes, that defeat was not a natural way to be, even for insects. The bees showed me that I had a choice how to live. I could collapse under the sadness of losing my parents, or I could keep going.

5

Big Sur Queen

1975—Summer

I started spending so much time in the eucalyptus tree that I began packing a lunch to take with me. If anyone noticed that I'd withdrawn from the family, nobody complained. I'm pretty sure my whereabouts went unnoticed. Except by one person.

I was more than halfway through a peanut butter and jelly sandwich when I could've sworn I heard an owl. I twisted on my perch, looking this way and that, but it was hard to see through the curtain of slender eucalyptus leaves rattling in the breeze.

"Hoo! Hoo!" Louder now. I shoved the last bite into my mouth and climbed down to a lower branch with a better view to scan the yard.

Grandpa. I spotted him hiding behind the wooden shed where he stored his bee equipment. He had his hands cupped in front of his mouth, directing his birdcall toward me. He was wearing his bee veil, and hooting through the mesh.

"I know that's you, Grandpa," I called down to him.

"Hoooow doooo yoooou know it's not an ooooowl?"

"I can see you."

He stepped into full view and looked up to the treetop. We eyeballed one another, waiting for the next chess move. Grandpa cleared his throat.

"Whatcha doin' up there?"

"Watching bees."

"Coming down anytime soon?"

"No."

Grandpa took his veil off and slowly folded it back down into a flat square. "That's a shame," he said.

I didn't answer, waiting to see where he was going with this.

"I needed someone to help me find the queen."

This was it! The invitation I'd been waiting for—to open up a beehive. It was the one thing, Grandpa knew, that could lure me out of the tree.

"Wait up!" I said. I scrabbled down the trunk so fast that the bark peeled underneath me in long, pink strips.

Grandpa kept more than one hundred hives sprinkled along the Big Sur coast. His largest bee yard was in a remote area at the foot of Garrapata Ridge accessible only by four-wheel drive, and even then sometimes he had to use a chain saw to cut through the occasional tree that fell across the road. Grandpa and a beekeeper friend owned a 160-acre piece of undeveloped Big Sur wilderness that he said was perfect for bees. Named for the Spanish word for *tick*, the

Garrapata Canyon caught full sun, was protected on either side by steep chaparral ridges and isolated from people. All the bees had to do was fly out of the hive and feast on California sagebrush all the way to the mountain peak, then float back down as their bodies grew heavy with nectar. The land was an all-you-can-eat buffet for bees, offering them a year-round menu of sage, eucalyptus and horsemint, while Garrapata Creek provided a clean source of water.

Each year his hives produced more than five hundred gallons of honey that he delivered to Big Sur customers, plus a couple local restaurants and a grocery store. He never advertised, because demand always outweighed supply. By fall, he ran out of honey and had to put anxious customers on a list for the following spring honey flow. I'd heard Grandpa tell stories about Big Sur at the dinner table, and it sounded like something out of one of my fairy tales, untamed and magical. I wasn't going to sit in a tree and miss my chance to finally go there.

Minutes later, I was bouncing on the passenger side of his work truck with my feet resting on a set of clanking metal toolboxes. It was a Chevy half-ton that farted like an old man and once upon a time used to be a glossy yellow, but now was weathered to the dull texture of chalk and pockmarked with rust. The odometer had flipped over to zero at least twice that he could remember before eventually it stopped spinning, and he attributed his truck's good health to a regimen of regular oil changes. The windshield was crusted with dead bugs and mustard-yellow dots of bee poop, which Grandpa couldn't remove with the wipers because they also had stopped working years ago. When rips appeared in the red vinyl bench seat, he covered them in duct tape; when he ran into something, he banged out the dings with a mallet. His truck was a handyman's flea market on wheels; everything he might need for beekeeping or plumbing jobs was tied to the contractor's rack, crammed in the back bed or jammed somewhere in the cab. The dashboard was packed several inches high with a nest of pipe fittings, grease-pencil stubs and rubber bands, opened mail and seed packets, and balled-up bits of beeswax. He used the hooks of the empty gun rack to hang his tattered work shirts, splattered with pipe dope.

I was wedged into a small space he cleared for me on the bench seat, separated from Grandpa by a barrier made of beekeeping magazines, his dented workman's lunch box and a green metal thermos. His dog, Rita, was in her usual spot, curled on an old pillowcase beneath his seat, safe from falling objects. The three of us literally clattered down the road, creating a jangling chorus every time we hit a bump and jostled Grandpa's collection of things that might come in handy someday.

When we turned off Carmel Valley Road south onto Highway 1 and entered Big Sur, nature woke up and suddenly started doing the can-can. Everywhere I looked, the jagged mountains were tumbling into the sea, like rockslides frozen in free fall—still yet dramatic at the same time. We navigated a thin, winding ribbon of road hundreds of feet above the exploding surf. I rolled the window down, and heard sea lions barking and waves booming into sea caves below. The spicy aroma of sage mixed with sea salt wafted into the truck. We dipped down into forests where the air dropped ten degrees and the massive redwood trees clustered together in tribal circles, then we burst back into the sun again. I twisted my head in every direction, trying not to miss a thing.

"There's one!" Grandpa said, pointing toward the ocean.

"One what?"

"Whales. Look for their spouts."

I squinted harder at the blueness.

"There it goes again!"

Grandpa was now driving with his head turned all the way to the right. I grabbed the armrest as he went around a tight left turn, but he stayed perfectly centered in his lane while he stared at the ocean. He'd driven this stretch of Highway 1 so many times he didn't need sight to navigate it.

"Where?" I scanned the horizon, but it looked just as blank as it had a second ago.

"It should come up again, right about there," he said, pointing farther south. "Sometimes you see two spouts, a little one next to a big one, then you know it's a mama whale with a calf."

As if on command, a white spray shot into the air from beneath the surface, and a beat later, a smaller one, just off to the right of the first.

"I saw it!" I yelped.

A turkey vulture circled effortlessly overhead on six-foot wings, its black feathers spread out at the wing tips like individual fingers. It was so huge it cast a shadow over the road as it passed above. I rolled the window down more, and the wind ruffled my hair as I looked up at the red of its head. We watched it glide above a cove with water the color of jade and kelp fronds waving on the surface.

"There's where you catch abalone," Grandpa said, pointing to the inlet.

"How?"

"You dive down with an abalone iron. You gotta get it under the shell quick, otherwise the abalone feel you doing something and clamp down on the rock."

"Does it taste good?"

"Yeah, if you hammer it first."

Sounded a little gross to me. I returned to whale spotting, but the ocean was a blank slate once again.

"See those two rocks?" he said, pointing to two triangular peaks jutting two stories high, less than twenty yards offshore. "I almost crashed right into them."

Grandpa unscrewed the cup on his thermos and held it out to me—my signal to fill it with scalding chicory coffee. Then he settled into one of his Cannery Row fishing stories. Grandpa used to fish alone in his own skiff for sardines and sell them to the canneries, but it was hard to compete with the large Italian family-run fishing crews, and he had to catch lots of fish to make any money. One day his friend Speedy Babcock told him there was more money for less effort in salmon.

"I had never fished salmon before, and Speedy said he'd teach me," he said.

They left Monterey for Santa Cruz in Speedy's twenty-eight-foot cabin cruiser, and caught thirty king salmon, about six hundred pounds of fish—a fortune. But on the way back, they got lost in the midnight fog.

"We couldn't see so we had to navigate by sound. The water sounds different in different spots along the coast, and he kept steering west, thinking we were turning into the Monterey harbor, but I could tell we were only at Point Lobos. He wouldn't listen to me. We argued until those rocks suddenly appeared and I wrestled the rudder from him. We almost lost everything by that much," he said, holding his thumb and finger an inch apart.

I asked Grandpa what happened after that.

"Never went fishing with that guy again," he said.

Grandpa slowed, put on his blinker and we turned left into the cool shade of Palo Colorado Road, lined with eucalyptus trees. On the corner was one of Big Sur's earliest homesteads, a three-story log cabin built in the late 1800s out of redwood slabs caulked together with lime, sand and horsehair. A sheep pasture encircled it, the lambs hopping straight in the air like grasshoppers. The ranch extended across Highway 1 to a stunning sea cliff pasture, where a herd of white-and-red Hereford cattle could stand close enough to the sea to feel the saltwater spray.

"My cousin Singy's place," Grandpa said, yanking his thumb at the cabin.

"Singy?"

"Yeah, everybody calls her that because as a girl she sang a lot."

"Are we going there?" I wanted to pet one of those lambs.

"Not today."

We continued on the narrow, winding road and soon the eucalyptus grove gave way to a cathedral of redwood trees. Palo Colorado Creek rippled along one side of the roadway, and sunlight filtered through the forest, making polka dots on tiny hillside cabins propped up on stilts over the creek. Staircases with too many steps to count zigzagged from the homes to the road. After a mile, Grandpa turned up a steep incline, driving through a tangle of greenery, the ivy-choked coyote brush and manzanita branches scraping our roof while the asphalt below became a dirt road the color of chalk rock. When we reached the plateau, we were in a meadow and we could see the sea again.

Grandpa stopped before a cattle gate that was secured by a locked chain. He reached into the glove compartment to pull out an enormous key ring just like the sort janitors carry. By the looks of it, everyone who had property in Big Sur had given him a key. He slid the keys around the ring, muttering to himself until he landed on the right one. He got out and popped the lock, unwrapped the chain and swung open the gate so we could pass through.

Grandpa shifted into four-wheel drive as we descended into Garrapata Canyon on an elevated wisp of a road with a severe drop-off on my side. It was barely wide enough for all four tires as the truck groaned around the tight switchbacks, bouncing over pits and boulders left by winter rains. Grandpa honked as he cranked the wheel around the turns just in case someone was coming the other way, and a few of the curves were so sharp that he had to do a three-point turn, backing up and turning, backing up and turning, before he could get the truck all the way around. One false move and we were toast. None of this seemed to bother Grandpa, who continued chatting away as rocks fell away from under his tires and skittered down the slope, but I couldn't bear to watch. I kept my eyes on the horizon, looking in the distance for the patch of ocean peeking through the V formed by the two canyon walls.

When we reached the bottom of the grade, pine needles cushioned us as we drove around fallen trees. Grandpa revved the engine and we drove right through Garrapata Creek, the water coming halfway up the tires. We got a tire stuck between two granite river rocks for a second, and the truck rocked back and forth as Grandpa tried to get momentum to propel us out of the divot. He seemed to be enjoying our predicament, wriggling his eyebrows at me as he punched the gas. Third time was a charm, and the truck bucked and splashed and got to the other side. We drove through more redwoods, and because the earth was wetter here, there were ferns and snarls of orange monkey flower encircling the trees.

Finally, we emerged from the tree grove into a small wildflower meadow, and Grandpa cut the engine. At one edge of the clearing was a city of vertical white beehives, each with a small cloud of black dots waggling before it. We stepped out of the cab to the sound of scrub jays complaining about our intrusion. The air smelled as clean as mouthwash—a minty mixture of bay leaves and sage and lemony horsemint. Grandpa opened his door, and Rita's long body shot out from underneath his seat, eager to hunt in the thicket.

"Get along, little doggie!" he sang after her. "Oh wait a minute," he said, watching her gather speed on six-inch legs, "I already *have* a long little doggie."

Grandpa laughed so hard his false teeth jiggled loose.

He'd lost his real teeth, he said, when they rotted and fell out in his twenties, despite regular brushing. Now he put his fake ones in a glass of water on the nightstand where they grinned at him as he slept.

He rifled through the stuff in the truck bed and tugged out two plastic hats with full brims. They looked like white pith helmets, with vents on the crown. He put mine on first, then slipped a yellow mesh veil over it so my head was covered all the way around, then cinched the netting in place with two long cords that he crossed over my chest, circled around my waist and then tied in the back. The hat was adult-sized and kept slipping over my eyes, but it was all he had.

He put his veil on, then lifted a burlap sack from the truck, fished inside for a dry cow patty, broke it apart and shoved the pieces into the can of the bee smoker. He lit the dung with a match, closed the lid and squeezed the bellows a few times to stoke the flame until white smoke coursed from the spout. As we approached the first hive, I saw a row of honeybees lined up at the slit at the base of the hive where the entrance was, beating their wings.

"Air-conditioning," Grandpa said.

Bees, he explained, always keep their hives about ninety-five degrees inside, no matter what the weather. In winter, you can put your hand on the outside of a hive and feel the heat radiating from within as they cluster together and shiver their wing muscles to generate warmth. In summer, bees gather on the landing board near the entrance and circulate air with their wings to cool the hive down. No matter where a hive is, whether in snow or triple-digit heat, it's always within a few degrees of ninety-five. How bees could regulate temperature so precisely without a thermometer was one of their biggest mysteries.

Grandpa handed me a metal tool, just like the one he always carried in his back pocket, with one flat end for scraping wax and a hook on the opposite end for lifting wooden honeycomb frames out of the hive.

"The bees glue the lid down," he said, showing me how to wedge the hive tool into the crack to pry the inner cover off. Bees, he explained, don't like cold drafts in their homes, so they make glue out of tree sap called *propolis*, and use it to seal any cracks in the hive. I mimicked his movements, and we each slid our tool under opposite sides of the inner cover. We popped it off, revealing a row of ten wooden slats underneath, each a removable rectangular frame of wax honeycomb resting on a groove cut into the box. The bees responded to the intrusion of sunlight with one quick, loud hum—a collective shout to warn the rest that something was happening to their home.

I peered closer, and noticed the bees were aligning themselves in rows in the empty spaces between the frames and peeking out to see what was going on. They wriggled their antennae, exploring the airspace where their honey pantry had been seconds before. The hive had a comforting smell of hot pancakes with butter and syrup. Grandpa reached in with bare hands and lifted out the first frame of honeycomb, which was blanketed on both sides by bees. They were a moving carpet, each an individual thread that together made one thing. Some went this way, others that way, bumping and crawling over one another but never causing injury or irritation.

Grandpa shook the frame over the hive to dislodge about half the bees, so I could see the honeycomb underneath. It was a masterpiece of mathematical symmetry. The interlocking hexagonal tubes were aligned in straight rows, every cell sharing one wall with six of its neighbors for economy of space and wax. To fight the laws of gravity, Grandpa explained, each honeycomb cell was slightly tilted upward a few degrees to keep the honey from spilling out. It was as if the bees knew that of the three shapes that can stack without creating wasted space—squares, equilateral triangles and hexagons—the hexagon uses the least amount of material for the largest storage room, thus saving on labor and supplies.

I reached with my fingertips to feel the geometry. The stacked configuration made the wax sturdy enough to hold several pounds of honey in one sheet of honeycomb, but the wax itself was pliable and crushed under my fingertip. Some of the cavities held gleaming honey, others small plugs of bright yellow and orange and reds where the bees had stored pollen grains. Grandpa turned the frame from side to side to examine it, bringing it so close to his face that his veil nearly brushed the bees.

"See the queen?" I asked.

Grandpa put the frame down on its side and propped it against another hive. The bees stayed on it, continuing to make their rounds on the honeycomb as if they didn't even realize that they had been ejected from their own home.

"Nah, this one is full of food, no place for her to lay an egg. She'll be in the middle somewhere, where it's warmer."

Some of the bees were now overflowing down the sides of the hive like a spreading stain. Instinctively, I took a step back.

"Okay, smoke 'em," Grandpa said.

I pointed the snout of the smoker over the remaining nine frames in the hive and squeezed the folding bellows once. One white cloud puffed out.

"Keep doing it. More. Lots more," Grandpa said.

I sent a storm of smoke clouds over the frames. The fumes had a wet cigar smell that tricked the bees into thinking their hive was on fire, sending them down into the hive to gobble honey before they fled their burning home. With full stomachs, Grandpa said, it made it harder for them to bend their bodies into stinging position.

When I had smoked most of the bees off the top bars of the hive, he lifted a second frame out. Grandpa worked bare-handed because he said he'd been stung so much it didn't bother him anymore. He swore all that venom prevented his joints from stiffening up with arthritis like Granny's.

He inspected two more frames, returned them to the box and lifted out another. Then he bent down on one knee and held the frame out to me so I could see.

"Look here, where I'm pointing."

I let out a small gasp. The queen was so obviously the queen. She was elegantly tapered, twice the size of all the other bees, and with longer legs that looked like they belonged on a spider. Her abdomen was so heavy with eggs that it dragged behind her as she walked.

Like bodyguards parting a crowd for a pop star, an entourage of attendant bees formed a protective circle around her and cleared a path as she moved. She rushed across the honeycomb like she was late for something. Her royalty was apparent in the way the other bees grew visibly excited when she came near, rushing up to caress her with their antennae, some even wrapping their forearms around her head in what looked like an embrace. Curiously, none of the bees turned its back on her. As she moved about, each new group of bees she approached rearranged themselves to face inward, even backing up before her to keep their eyes and antennae focused on her every move. The only word for their behavior was *worship.*

"Why do they touch her like that?"

"They are gathering her special scent and passing it to the rest of the bees," Grandpa said. "That's how they know which hive is theirs. Every queen has her own smell. Her daughters never forget it."

It's true mothers have an aroma. Mine smelled like Charlie perfume and Vantage cigarettes, mixed with the faint musk of other people's clothes from the church thrift shop. It was a unique scent that I recognized instantly whenever I climbed into bed. I thought of Mom at that moment, passing the hours in bed. I wished she could see this queen, how an insect was so perfectly designed to be a mother, how the queen was the heartbeat of a whole stunning society operating right under our noses. There were so many fascinating things happening outside Mom's four walls, but she was missing out on all of it. Her days came and went without little miracles like this to lift her spirit.

The queen padded along the honeycomb with the impatience of the very pregnant. She seemed to be weary of all the attention, refusing to slow down for every bee that wanted to touch her as she single-mindedly searched for something. Every few steps, she ducked her head into one of the honeycomb cells, then retreated. She checked chamber after chamber, hunting.

I asked Grandpa what she was looking for.

"A good spot to lay an egg," he whispered. "Gotta be clean and well built. Can't have an egg already in there."

The queen squeezed her body inside the honeycomb to inspect, and all that remained visible was her butt sticking out. She was picky about her nursery, but finally she found a space to her liking and backed her abdomen into it. As the queen crouched there for a second, her attendant bees came in close as if to tell her a secret. The queen then did a little push-up with her legs and exited as her admirers backed up to give her room. I peered at the hexagon cell where she'd just been and spied a white pin inside, like a miniature grain of rice, standing on end perfectly centered on the back wall. Two of her attendants dipped their heads into the cell to verify her work. I had never seen anything born before and realized I'd just seen my first miracle.

"Is she going to do it again?" I whispered.

"About a thousand times a day," he whispered back.

Grandpa stood back up and gingerly slid the frame with the queen back into the hive, being extra careful not to squish her. He stacked the hive back together and closed it, then moved to the next one. He wedged his hive tool under the lid of the second hive and broke the sticky propolis seal, then twisted the top box off and set it on the ground, his cheeks puffing with effort.

What struck me most about the queen was how many children she had. That seemed like an impossible number for one mother to handle.

"Hey, Grandpa?"

"Hmmmm?"

"How can one queen take care of that many bees?"

He slid his tool into his back pocket, and pushed his mesh veil above his eyes and perched it on his forehead, so he had a clearer view of me.

"All the bees take care of each other. A hive is like a factory. All the bees have different jobs so they share the work."

I gave him a sideways look and crossed my arms skeptically. Grandpa put the smoker on the lowered tailgate of his truck, away from the dry weeds. He squatted in front of a hive and waved me to come closer. He pointed at a handful of bees that were standing in a cluster at the entrance with their backsides facing outward, all ferociously beating their wings.

"Their job is to air-condition the hive," he said. Then he pointed at another bee on the landing board.

"Now watch what this one does."

The bee marched left, then right, then left again, as if it couldn't make up its mind where it wanted to go. Just then a second bee landed nearby, and the pacing bee scurried over, crouched defensively and blocked it from entering the hive. The first bee circled the new arrival, tapping it with its antennae, then stepped aside and let it continue inside.

"Guard bee," Grandpa said. "Making sure no strange bees enter the hive."

I was stunned. Until now, besides the queen and the stocky male drones, all bees looked identical to me. What had seemed like aimless crawling of thousands of bees now snapped into tight organization, once I understood that the way to see bees was to watch their behavior. I pointed at the bees landing at the entrance.

"What kind are those?"

"Field bees. They bring nectar and pollen. House bees that stay inside the hive will take it from them and store it."

"Can I see?"

He reached into the hive and lifted out a honeycomb frame blanketed with bees. I pointed at a bee that had its head buried in one of the cells, and asked if that bee was storing honey. He brought the frame closer to his face and blew on the bee gently, and it backed out of the cell so he could see what was inside.

"Nope. That was a nurse bee feeding a baby." He lowered the frame and pointed. Inside the cell was a small white grub.

The more Grandpa taught me, the more excited I became. I wanted to understand everything that the bees were doing, to be able to read them the way he could. Because when I let myself get lost in a beehive, my mind could stop spinning. I was able to slow down and relax with the task of simply paying attention. Serenity came as I shifted my worried mind to the bees and their behaviors. I felt a comforting assurance that there was hidden life all around me, and that made my own problems seem smaller somehow.

I learned that some bees are wax makers, others are builders that construct the honeycomb, and there are even undertaker bees that remove the dead, flying out of the hive with bee corpses in their clutches and dropping them far from the hive. Grandpa explained that a bee will have many different jobs during its life, but every bee's first job is janitor, cleaning the debris out of the honeycomb and polishing the cells so they can be reused for storing honey or laying eggs. A bee promotes upward through the various in-house jobs, nursing the babies and curing nectar into honey, until reaching its final job foraging for food outside the hive. Now it made sense why the queen could lay so many eggs a day. She had a massive day care system in place. Her only job was to drop eggs in cells.

"The queen can't even feed herself," Grandpa said. "Those bees you saw in a circle around the queen? That's her royal court. They bring her water drops when she is thirsty, food when she is hungry. They keep her warm at night, and they even clean up her poop!"

"What happens if the queen dies?"

"The bees will make a new one."

You can't just make your own mom. No animal had ever done that on one of our nature programs. I wasn't buying it.

"That's impossible," I said.

"Not for bees," Grandpa said. As soon as the bees sense that their queen is failing or missing, he said, they select a handful of eggs and start feeding them royal jelly—a milky superfood the nurse bees produce from glands in their head. It's full of vitamins, and a steady diet of it will make a regular worker bee larva start to develop into a large queen.

The bees build protective wax chambers for the incubating queens that look like unshelled peanuts dangling from the honeycomb. Wait a couple weeks, and the tip of her birthing chamber turns papery and thin. She chews her way out and, Presto! New mom.

"Bees are very smart, but most people don't see it," he said.

"But you said a hive can only have one queen," I countered.

A colony raises more queens than they need for insurance, he explained. The first virgin queen to emerge races to tear open the other queen cells and stings her rivals to death. Grandpa wiggled his eyebrows at me for dramatic effect.

"Really?" I whispered. Grandpa had convinced me that bees were gentle, and now they seemed capable of terrible brutality. I bit my lower lip, unsure of what to think.

"Why would I kid you?" Grandpa said. "You can even hear the queen fights. They let out a battle cry that sounds like a duck quacking. Yeah, it goes like this—*waaaah... waaaah...waaaah...wah-wah-wah.*"

It was an astonishing thought, to replace your mother. What if humans could do that? I imagined a store that sold mothers, and I all had to do was walk down the aisles of ladies packaged in Barbie boxes and choose. What kind of mother would I pick? Mine would have long, shiny blond hair and a name like Gloria. She would wear pantyhose that came in those plastic eggs, and her high heels would click-click-click when she walked. She would come to my classroom and help all the kids with their art projects, and put Snoopy Band-Aids on my knees when I fell down. I imagined us driving in a convertible, and she'd have a long yellow scarf that would blow behind her. She would always let me pick the song on the radio, and take me to a drive-through for burgers and fries whenever I wanted.

Grandpa tapped my shoulder and my daydream popped. He had another frame in his hands, but on this one the honeycomb in the center of the frame was not orange with honey, but the cells were sealed shut with dark wax the color of a brown paper bag. He pointed again, and at the tip of his finger I saw two small antennae poking through a very small hole in the brown wax, where a bee was emerging. From behind the wax, the bee pushed and bit at the hole until it was large enough to poke its head through. The fuzz on its head was bright butter yellow, and matted down like it was wet. Its antennae swiveled as it explored the outside. Several bees ran over to touch the newcomer, and it startled and retreated back into the cell. Grandpa pulled a dry weed out of the ground and used the tip to pull the wax away from the cell opening, giving the baby bee a clear path to come out. It ambled out on wobbly legs, stood for a moment and stretched its wings. The newborn immediately began begging for food from passing bees, and within seconds an older bee stopped and linked tongues with it to pass honey, and the baby ate greedily.

I had no idea there were so many things going on inside a beehive. Grandpa examined all thirty of his colonies, and each one was different from the next. Some hives were swelling with bees, and others looked lonely for company. Some had cranky bees that ran over the comb like they had the heebie-jeebies, and some had sweet bees that ignored us as we inspected. Some were busy making queens, and others were hoarding pollen. Some colonies made wacky wax sculptures inside, and others formed precision sheets of straight comb. One hive even had two queens, which while rare sometimes happened when the queens decided to be friends, which made me relax a little bit about the power struggles over the throne. I was beginning to see that every hive had its own mind, and a good beekeeper keeps track of which hive needs what kind of attention.

The sun had settled down to the waterline by the time Grandpa had finished, and the beehives were making long shadow bars on the grass. As we walked back to the truck, two quail parents heard us coming and hustled their brood behind the safety of the sagebrush, the babies scurrying like cotton balls blown by the wind. Once we were settled back in the truck, he reached down under his seat to see if Rita licked his fingers. Satisfied she was onboard, he put the truck in gear and we bounced back up the pitted dirt road, but this time I knew Grandpa had it under control.

"I like it here," I said.

"Yeah, me, too. A person can think in Big Sur."

I understood exactly what he meant. I'd just spent the last several worry-free hours thinking about nothing but bees.

Once we reached the smooth pavement again, Grandpa pointed south down the Coast Highway and told me that when he was in fifth or sixth grade, every day he'd hike five miles up Bixby Canyon to work on the Chapman Ranch with the Trotter brothers. The siblings were teenagers, already huge for their age, and they taught Grandpa how to haul hay, split redwood into timber, brand cattle and shear sheep. Eventually, they were the ones who taught Grandpa how to be a plumber. Grandpa paused his story for a moment as if remembering something, then he began explaining the proper way to pull a lamb out of a ewe.

"If it's coming out backward, you have to reach in and grab what you can and turn it around." His voice was grave, as if what he was sharing with me would one day save my life. I didn't have the heart to tell him that I would never, ever, stick my hand inside an animal of any kind, for any reason.

I rolled my window down to let in the salt air. The mountains were turning dusky purple in the fading light, and a red-tailed hawk tracked our passing truck from atop a telephone pole. I felt oddly content, as if nothing bad could happen to me in Big Sur. I had managed to spend a whole day exploring inside a hive, too absorbed in learning about the bees to feel a pang of sadness. Big Sur was like a secret trapdoor to a pleasant dream.

Watching the queen bee work tirelessly for her family, and her children jostle to take care of her, helped me feel a little less bad about the family I had lost. It reassured me that motherhood is a natural part of nature, even among the tiniest of creatures, so maybe there was hope yet that Mom would come back to me. Even though the bees left the hive every day, they *always* came back. There was never any doubt that a bee had any other purpose than to be with its family. The hive was predictable, and that was reassuring. It was a family that never quit.

6

The Beekeeper

1975—Fall

When Granny took me to the church thrift shop to buy school clothes, I knew that we were going to stay in California for good. I accepted this with a child's surrender, the feeling of floating down a river on a boat steered by others, watching the turns of my life come into view with a quiet detachment. As was custom in my family, there was no conversation to explain why our visit had become permanent. On the one hand I was thrilled to finally meet kids my age, but I was also sad to forfeit a private hope that one day we'd return to Rhode Island and be a family again.

The thrift shop was in an attic above the church, accessible by a stairway behind the sanctuary. The room was musty, with angles of light from a few small windows at the roofline illuminating the dust motes floating in the air. Granny let me pick out a shirt, and I chose a white button-down, short-sleeved blouse with vertical green stripes. Looking closer, I saw the stripes were actually columns of Girl Scout emblems—little symbols that looked like four-leaf clovers. I couldn't believe my luck—a real piece of the official Girl Scout uniform. Granny pushed aside the hangers on a crowded circular rack and tugged out an ankle-length padded skirt with a patchwork pattern of gingham and calico squares. It looked very much like she wanted me to wear a quilt to school.

"This is respectable," she said, holding it aloft.

I wasn't sure what she meant, but I knew that when Granny made up her mind about something, the proper response was compliance. My shirt and Granny's skirt together were a train wreck of an outfit—*Little House on the Prairie* on the bottom and wayward troop leader on top, but this was what I chose for my first day of school, with sneakers.

There was no fanfare on my first day at Tularcitos Elementary. Mom remained in bed, Grandpa left before the sun for a plumbing job down the coast, and Granny whisked Matthew and me out the door with her. Now that the school year had started, she had to leave earlier in the morning to get us to day care at a lady's house in the village, before she drove to Carmel to ready the day's lessons before her fifth-graders arrived. I had breakfast with the day care kids, then walked myself to school, using a shortcut through the dirt airport.

It was common in the seventies to see kids walking everywhere by themselves in Carmel Valley. Crime was rare, and the village was so small that everyone knew which kid belonged to whom and kept a collective eye on our whereabouts. Our neighborhood was scarred with footpaths worn through fields and behind homes where kids had created their own transportation networks linking the convenience store to the community pool, the library to the baseball field. So the plan Granny laid out for me was to walk to Tularcitos each morning and back to day care each afternoon, where I'd wait with Matthew until she could retrieve us. I became a latchkey kid without a key.

On the first day of school, I kept to the edge of the road, inhaling the licorice scent of the wild anise bushes and turning to look over my shoulder every now and then to keep a lookout for the occasional car. The street was sleepy and deserted in the early morning; even the neighborhood dogs were still snoozing as the first rays of sun warmed their bellies. I passed a horse corral where two ponies lifted their heads expectantly. Normally, I would have stopped to feed them tufts of green grass through the fence, but this time I hurried along so I wouldn't be late for my first day of school. I cut through the airport, and finally reached the white ranch house next door to the school with the weathered wagon wheels leaning against the porch. That's when I heard the glorious cacophony of children's voices emanating from the schoolyard. I stood there for a moment, just listening to the lovely echo of potential friends.

The centerpiece of the schoolyard was a two-story jungle gym made from old telephone poles, fashioned into two fortresses connected by a chain suspension bridge that swayed perilously when kids ran across it. We were guaranteed to get splinters every time we climbed the structure, and we scorched our bums on the metal slides that got griddle hot in the sun.

When I reached the playground, boys and girls were streaking across the swaying footbridge, hopping over the missing slats and somehow staying upright as it lurched side to side, chasing each other with the delight of renewed rivalries. Other kids zipped down steep metal slides, screaming for those below to get out of their way. Boys crawled on their bellies like soldiers through tunnels created by industrial-sized clay pipes that were half buried in the sand. Girls swung from the monkey bars, their hair flapping behind them as they expertly flew from ring to ring down the line, the metal chiming with each catch and release. In another corner of the sand lot, girls were executing gymnastic tricks on the horizontal bars. One girl in pigtails sat on a crossbar about six feet above the sand as a cluster of her friends chanted from below, "Dead man's drop! Dead man's drop!" I watched the girl let go and fall backward, swing around the bar on the back of her knees and somersault into the air to land on her feet.

I felt a zing in my fingertips. I joined the river of kids and let the adults with clipboards guide me to my classroom. Students were gathered on the floor before the teacher, who was taking attendance. As I approached, I heard snickering and immediately flushed. I was disastrously overdressed. The girls wore Ditto jeans with hearts or rainbows sewn on the back pockets. The boys wore Levi's or corduroy shorts, and T-shirts with surf logos or Adidas stripes. I was woefully out of place with the padded fabric of my skirt that held its dome shape when I walked, so it almost looked like I had some sort of petticoat underneath. This was what happened when you let old people dress you. Granny chose the kind of old-timey clothes she used to wear when she was young.

The girl sitting next to me had hair so blond it was almost white, and in a certain light I thought I detected a green tint to it. She had her hair cut in a bowl shape, like the ice skater Dorothy Hamill. She wore a pink satin jacket.

She told me her name was Hallie.

"Why's your hair green?" I asked.

She frowned.

"The pool turns it green."

"You have a pool?"

"Mmm-hmm. I have a trampoline, too."

She probably had her own room. With a television. During recess, I followed her to an area of the yard reserved for kickball. I was one of the last players picked for a team, and when my turn at home plate came, my ankle-length skirt didn't allow me enough room to swing my leg and get a good kick. I had to take little doll steps to run the bases, and invariably was thrown out every time. Hallie was such a good athlete that the fielders backed up each time her turn came up. She flew around the bases with long strides just like one of the boys, pumping her arms and exhaling in powerful bursts. She was a marvel. The bell rang to go back to class, and I fell into step with her.

"You're really good," I said.

"It's easier if you wear pants."

I promised her that I would.

That night I ditched the skirt to a far corner of the bedroom closet, hiding it behind the winter coats. I needed to be more careful not to let Granny embarrass me again. I vowed to pay more attention to my classmates, and do what they did so I would fit in. I observed them with an anthropologist's eye, looking for clues to what I was supposed to want, how I was supposed to behave. I eavesdropped on their conversations about Disneyland, the zoo and McDonald's. I copied their slang, memorized the pop songs they sang. I cataloged the items they pulled out of their lunch bags—juice that came in silver pouches, sticks of cheese they pulled apart in strings, flat fruit that they peeled from cellophane. Hallie showed me how to twist her Oreo cookies apart and lick the frosting first. It tasted marvelous, like ice cream that didn't need to be kept frozen. But no matter how much I wheedled inside the Safeway store every Saturday morning, Granny refused to buy such ridiculous things. Not only did she not understand what they were, they were outrageously expensive. Mom's lack of income meant I was entitled to free government lunch at school. In the court of Granny, there was no arguing against free.

But sometimes free comes with a price. In the cafeteria I stood in the special lunch line, which everyone knew was for the kids whose families didn't have enough money for groceries. I envied the students with their Mom-made sack lunches, and listened to the daily frenzy of their bartering as they swapped gummy bears and peanut butter saltines, and sandwiches on white pillowy bread with the crusts cut off. Every day I got a hot meal in an aluminum tray sealed with foil, and no matter what was inside, it always smelled like boiled potatoes and was completely leached of flavor. No one wanted to trade for gray broccoli and limp fish sticks, so I started spending lunch and the recess period that followed inside the classroom with my smelly food, flipping through Dick and Jane books. My teacher urged me to play outside, but I refused so often that eventually she stopped trying. She and I ate together indoors, she working at her desk and me on a beanbag chair, satisfied by silence between us.

I scored low that year on the Social & Emotional Growth section of my progress report:

Works very hard in classroom; I often have to "throw her out" at recess. Complains occasionally that she is bored—both at school sometimes and after school. Have encouraged her to exchange phone numbers with classmates and get together with them.

I gave my progress report to Granny, along with her cocktail. She sipped her drink and glanced at the report, told me I was doing fine in school, then tossed the paper in the fireplace, where Grandpa was jabbing the poker into the orange flames. He made a fire at least once a week, even in warm weather. Our fireplace wasn't only used to heat the house; it was a tool to get rid of stuff. There was no recycling program, so my grandparents tossed newspapers and milk cartons, and old rags and magazines, and Kleenex and the occasional Sears catalog into the flames. Granny looked content as she watched my report card seize in on itself and turn to ash. She raised her glass as if to make a toast. "Who needs friends? Hell is other people, if you ask me," she said.

I didn't exchange phone numbers with anybody. The other students didn't invite me to their homes, but I didn't dare invite anyone over to our house, either. There was a secret behind the closed bedroom door in our house. I didn't want to keep Mom hidden, but I didn't want to explain to a classmate why she wouldn't come out of the room. I'm not sure I could give a reason, anyway. I already felt like an outsider at school for having grandparents instead of parents, and the inexplicability of Mom would only amplify my weirdness.

When I came to bed later that night, I found Mom asleep on her back, with a big red book splayed across her chest. Linda Goodman's *Sun Signs.* Recently Mom had discovered astrology, and pored over books Granny fetched from the library, looking for the cosmic explanation for her divorce. I gently slid the book from under her hand, trying not to wake her. She jerked awake and her eyes flipped open. I saw her eyes register the room, then she relaxed back into the pillows and reached for me. "It's okay, come on."

I got under the covers and tucked my bum into her belly, and she drew up her legs and pulled me into our nightly position.

"You're a good girl," she said. "For an Aries."

Mom had divided all the signs into good and bad people. I was a ram, which she explained is sort of a self-centered person, but fun to be around, and deep down, good. But a Taurus was the best, Mom said, because she, Granny and Matthew were all that one. But Grandpa was a ram, so I was happy.

"Mom?"

"Hmmmm."

"Halloween's coming."

The black and orange decorations were already up at school. All the classrooms were planning parties, and the only thing anyone could talk about was their costume. I wanted to be Dorothy from *The Wizard of Oz*, and I asked Mom if she'd sew me a dress. She'd made me a Raggedy Ann costume in Rhode Island, and it was perfect.

"I just can't," she said. "Ask your granny."

Granny was no help. She didn't sew, plus Halloween was just another way to spoil children, she explained. They didn't celebrate Halloween when she was a kid, and she turned out just fine. I tried to tell her that Halloween was the most important day of elementary school. The one day when you could eat all the sugar you wanted and blame any bad behavior on your persona. The teachers were promising costume contests, and we were going to carve pumpkins. If I didn't have a costume, I wouldn't be participating, so I might as well stay home. Granny harrumphed and reminded me I don't make the rules in her house.

It didn't occur to me to ask Grandpa for help; I couldn't imagine him with a needle and thread in his burly paws. Even if I had come to him with my problem, it would have gotten back to Granny, who had already told me to stop bothering her about a costume.

When I awoke on October 31, there still was no plan. Grandpa had already left for a plumbing job in Big Sur, and I found Granny in the kitchen rushing to empty the contents of Grandpa's wooden shoeshine box onto the counter.

"Sit here on the stool," Granny said.

I obeyed. She twisted the lid off a round tin of brown shoe polish, dabbed her finger in and smeared it on my forehead.

"Now hold still," she said, tipping my chin toward her.

"What are you doing?"

"Making your costume," she said, adding black around my eyes. She quickly covered my whole face and part of my neck. Next she grabbed one of Rita's brown flea collars from the broom closet, and buckled it around my neck.

"Wait here," she said.

I heard her opening drawers in her bedroom. She returned with a balled-up pair of beige pantyhose. She flicked her wrist and they unfurled, then she stretched the elastic waist over my head, and put all my hair inside. The legs flopped down to touch my shoulders. Last, she clipped one of Rita's thin leashes to the flea collar, then handed me the other end.

"Okay, that oughta do it," she said, stepping back to check her work.

Granny followed me to the bathroom to go have a look. I stood before the mirror and gasped. It looked like I had been horribly burned, just the whites of my eyes peering out from a chocolate brown face with black lines drawn across my forehead, and dark circles around my eyes. There was a black triangle on the tip of my nose, and whiskers drawn on my cheeks. I looked like someone with leathered skin from spending too much time outdoors, wandering around with pantyhose on their head. My mouth fell open, and I touched the greasy paint to see if I was still underneath.

"You're a basset hound!"

"A basket what?" My voice came out in a whisper.

"A dog, a hound dog."

She'd read a magazine article about how to make Halloween costumes with ordinary household items.

"I look dumb," I protested.

"I'll tell you what's dumb," she said. "There are children in other countries who are starving, and you're worried about a Halloween costume."

It was done. There was no more discussing it. I slumped on my daily walk to school, carrying my own leash. The shoe polish had a strong petroleum odor and was making me slightly woozy. On the playground, I parted a sea of perplexed princesses and superheroes as they struggled to figure out what I was.

Hallie was dressed as a ballerina with a red tutu over her gymnastics leotard, and pink ballet slippers with ribbons that crossed up her calves. She covered the sun from her eyes and squinted to get a better look at me.

"Why do you have underwear on your head?"

"Those are ears."

Hallie's forehead wrinkled in confusion.

"I'm a hound dog."

I kept my eyes on my shoes. "Granny did it. It's not any good."

Hallie took the leash from my hand.

"You can be my dog," she said. "If anyone says anything, you attack them on my command."

The beautiful part of the plan was that as her dog, I could remain mute and not have to answer any questions about my costume. Hallie spoke for me, explaining that every ballerina has a guard dog, end of story. When our teacher took a classroom picture in the sandbox, Hallie held my leash, and I knelt down at her feet, her loyal pet. Our plan worked, and I kept my dog face on until I could no longer stand the fumes. I scrubbed the shoe polish off in the bathroom using industrial pink powder soap and scratchy brown paper towels. Lastly, I yanked the pantyhose from my head and dropped them in the trash can.

No matter how much trouble I was having fitting in, I still liked school itself. I embraced the routine of it, the bells that set the parameters around art projects and recess and story time, putting a purpose to my days. Every day I came home with a story for Grandpa about what I'd learned, and he gave me encouragement to keep trying to make friends, reminding me that it takes time to find the right people who you feel comfortable with. When I told him what had happened on Halloween, he gave me two pieces of advice: keep Hallie as a friend for life, and next year put on his bee veil and go to school as a beekeeper. I couldn't believe I hadn't thought of it.

Our teachers were hippies and nonconformists who freely embellished on the curriculum. One teacher taught us how to shape and fire clay pots. Another tested our extrasensory perception, drawing a symbol on a piece of paper and challenging us to harness our ESP power to duplicate what he'd drawn on our own papers. For some reason this exercise had to be done on the soccer field, where we stood in a circle around the teacher, sketch pads in hand, and tried to read his mind. I learned from a science experiment that cola will rot my bones. A teacher put three Dixie cups of cola in the window, and inside one she dropped a chicken bone, another a nail and the third a penny. We monitored the daily deterioration of each item in a logbook. When the chicken bone disappeared first, within a month, I promised almighty God that I would never drink soda again.

I couldn't wait to get to school and find out what the day's discovery would be. I responded to my teachers' attention with a gratitude I didn't comprehend beyond the feeling that I wanted to please them, to memorize everything they told me, and show them how good I could be.

One day, the surprise was a new music teacher. The first time I walked into his classroom, Mr. Noakes was sitting with his knees apart on a metal stool, strumming a guitar, like he was waiting for a bus instead of a roomful of kindergarteners. He was bamboo-thin and looked too young to be a teacher, in jeans and tan suede Wallabee boots with flat rubber soles, periodically flipping his long brown bangs out of his eyes so he could locate the right frets. He rumbled up to school in his VW van on Wednesdays only, for the last hour of school, and his class became the highlight of my week. On music days he'd prop open his door, lower the needle on the vinyl and let the lyrics draw us in. "Bad, Bad Leroy Brown" would filter down the breezeway, we'd cock our ears, drop our pencils and join the Pied Piper parade to the music room. Mr. Noakes didn't play that "Puff the Magic Dragon" happy music designed to calm us down—he let us listen to real songs from the radio.

When Mr. Noakes let us choose musical instruments, most of the girls gravitated toward the twinkly-sounding flutes or xylophones while I jostled the boys for the drums. He let us make noise, and he never blew whistles at us like some of the other teachers, nor did he keep a running tally of demerits for bad behavior on the chalkboard. He saw it as his personal mission to instill excellent musical taste in our malleable brains, and played records for us from his personal collection. One day he flipped through his milk crate, pulled out an album cover and held it up for us to see.

"Does anyone know who these guys are?"

I recognized the four Beatles in a crosswalk and instantly froze. That was Dad's music. Suddenly, my skin felt clammy and the floor beneath me tilted. Mr. Noakes was still clutching the *Abbey Road* record, his eyebrows arched, waiting to see whether anyone could identify it. I raised my hand, along with one other boy.

"Only two of you?"

Mr. Noakes scanned the room, savoring the moment before blowing our innocent minds. He looked giddy as he approached the record player and worshipfully slid the black disc out of the sleeve, being careful to keep his fingertips to the outer edge of the vinyl, and lowered it on the turntable.

This could not be happening. The Beatles were private between Dad and me, not something everyone could just have for free. Playing it in front of the whole class would be like *prying into my life*, and Mr. Noakes had no right to do this to me. I helplessly watched him put the needle on the record, knowing that a terrible secret was going to flood out of me, something that I was not allowed to talk about at home, something that I was ashamed of, something that would separate me further than I already was from my classmates. I looked to the door and wondered if I could make a run for it.

The first notes of "Maxwell's Silver Hammer" swelled from the speakers, took hold of my body and shook it. I felt heat ripple up from my stomach, rise up my throat and collect behind my eyes. I didn't hear Paul McCartney; I heard Dad's voice telling me to go to sleep, to finish all my peas, promising he would always be my dad. It was as if he had materialized in the room, but when I tried to look into his face, it kept slipping out of focus, like he had stepped behind an opaque screen. I panicked, trying to remember what he looked like. All I had left of him was my memory, and I was starting to lose that, too. I looked around and saw that the students were engrossed in the strange, perky song about three murders, laughing and pretending to clobber one another with a hammer. I would never have pure joy like they did. I hated them for being so effortlessly happy.

I could feel tears gathering and willed them to go away. I couldn't afford to add a meltdown to my growing list of social transgressions. I squeezed my eyes closed and hummed, trying to block out the song. When that didn't work, I put my forehead on my knees so my jeans would absorb my tears. A few sobs slipped out, and I tried to make them sound like hiccups. My chest heaved, and snot ran down my upper lip. By the time the song ended, the only sound in the room was my weeping.

Mr. Noakes hastily dismissed the class, and I stayed curled in a tight ball. When the room was empty, he knelt next to me.

"What's wrong?"

The sound of a man's voice only made me shudder harder.

"My dad..." was all I could get out.

"Oh man," Mr. Noakes said under his breath. "Don't move. I'll get the nurse."

She appeared in the room, huffing from a run. I let her lift me off the floor and wrap me in her thick arms, and I melted into her big bosom. Hugging her was like burying myself under the bedcovers, and I stayed there until I'd stopped snuffing. She held my hand and walked me to her office, where I sat on her cot and tried to tell her why I was upset. It was very hard to explain.

"My dad," I repeated.

She handed me a tissue. "Where is he?"

"Rhode Island."

She blew out her cheeks and paused a second before pulling out a metal file drawer. She rifled through manila folders and lifted one out. She held it open in one hand, and asked me her next question without looking up.

"Do you live with your mom?"

"Yes. No… I live at Granny's."

She tilted her head, like she was trying to figure out what I was not telling her.

"Who should I call to pick you up?"

I told her that nobody picks me up.

"I walk home," I said, pointing east.

She pulled a pen out of a cup on her desk and scribbled a phone number on a notepad, tore it off and handed it to me.

"Give this to your grandmother when you get home, and tell her to call me," she said. I nodded.

"Do you need to rest awhile before you're ready to go?"

I declined. I'd had enough of this day and was ready for it to end. When I gave the note to Granny, I was too scared to tell her the truth, so I told her I didn't know why the nurse wanted her to call. Granny didn't press further, and I was happy to not talk about it.

The following Wednesday, when it was time to go to the music room, my teacher told me to stay behind in class. When the students were gone, she placed before me a new set of watercolors and a pad of paper. She poured some water in a cup and handed me a brush. I stared at the blank page for a moment, then painted the first thing that came to mind. Six legs, four wings, three body sections, five eyes, two antennae. A stinger.

Over the next several weeks, I continued to paint while my classmates went to music class. I missed playing the drums, and even though my teacher said I could return whenever I felt ready, I never felt ready. Now the kids tiptoed around me as if I were fragile, which was at least a step up from being ignored. My painting progressed. I made pictures of pretty houses with curtains in the windows, and stick trees topped by big green balls of foliage. I painted cats and bees and flowers. I brought them all home to Grandpa, who carefully admired each one and taped them to the walls of his "office"—an unfinished room off the carport where he kept an old Western desk and boxes of plumbing supplies.

One afternoon I found him in the carport, stomping on aluminum cans and then finishing the job with a sledgehammer, pounding them into flat discs. He gripped the handle with both hands, raised the sledgehammer upside down and smashed the cans with the top of the head. He was tossing squished cans into a cardboard box in the back of his truck when he saw me.

"You get good money for these at the scrap place," he said. "Five cents."

By the looks of his pile, I guessed he would make a small fortune. There were hundreds of cans littering the ground. His white T-shirt was worn so thin there were holes in it, and the bottoms of his pant legs were wet from remnants that had squirted out of the cans. He was wearing leather boots, the toe of the left one encircled in duct tape to mend a hole. Bits of food were stuck in his mustache.

"Whatsa matta you?" he said, noting my long face.

I told him we were going to have a special day at school. That the dads were going to come and each say a little something to the class about their jobs. That I wasn't going to go because I didn't have a dad to bring.

"I see," said Grandpa, taking a long pull on his beer. He burped with gusto. "Pardon me."

He dropped his beer can on the ground and flattened it. Then he set the sledgehammer before me. "Want to do one?"

I grasped the handle and strained, able to lift it only a few inches. I stepped my feet a little wider and let the full weight of the sledgehammer come down, and the can yielded with a satisfying crunch. I felt powerful and suddenly found a hidden reservoir of energy. I hit the can again, and again and again, losing myself as the beer foamed out of it, feeling a little better with each blow. When I finally looked up, Grandpa was staring at me. He asked if I was working on any new art projects, and I mentioned we were learning papier-mâché.

Grandpa lifted an eyebrow. "What're you making?"

Grandpa visiting us in Newport, Rhode Island, 1974. I am 4.

Grandpa, Matthew and me in Newport, Rhode Island, summer 1974.

Mom and me shortly after my first birthday in Newport, Rhode Island, 1971.

My brother, Matthew, and me in 1975, just after we moved into our grandparents' house in Carmel Valley. Behind us is the eucalyptus tree I would climb to get as close as I could to the honeybees.

Grandpa, Matthew and me in Carmel Valley, 1976. I'm holding Harold and Grandpa is holding Rita.

My grandparents' little red house on Via Contenta in Carmel Valley.

Every spring Grandpa caught honeybee swarms,
like this one he scooped out of a tree in Pacific Grove in 1994.

Granny and Grandpa celebrating
their 25th wedding anniversary in 1988.

The honey bus. Originally built in 1951 for the fleet at the Fort Ord military base, Grandpa bought the bus from a Big Sur friend in 1963, tore out the seats and built a honey bottling factory inside.
(Photo by Jenn Jackson)

The left side of the honey bus. Wooden pallets in the foreground served as stairs to the back door. The small tower in the middle ground is the propane-heated water boiler Grandpa designed to send steam in a hose to the blade of the "hot knife" he used to cut open honeycomb.
(Photo by Jenn Jackson)

Inside the honey bus. The honey spinner is in the foreground; behind it are two honey holding tanks with cheesecloth strainers. The Wesson Oil cans on the right were used to sell honey to customers who wanted large quantities. (**Photo by Jenn Jackson**)

The honey spinner, or extractor, used to spin honey out of uncapped honeycomb using centrifugal force. The pulleys were powered by a motor Grandpa cribbed from a lawn mower (hidden under the plastic). (**Photo by Jenn Jackson**)

Grandpa lifting a frame of honey inside the bus, 1986.

A honeybee parting its mandibles to suck up a drop of spilled honey through its straw-like proboscis.

Grandpa slicing open wax honeycomb before putting the frame in the spinner, 1986.

The Pacific Ocean peeking through the walls of Garrapata Canyon in Big Sur, on the drive to Grandpa's apiary.

A typical cabin in Palo Colorado Canyon, Big Sur, where Grandpa made the bulk of his honey deliveries.

The dicey mountain drive to Grandpa's beehives in Garrapata Canyon, Big Sur.

Grandpa in his Garrapata bee yard, mid-1960s.

A family party at the Grimes Ranch, one of Big Sur's earliest homesteads. Grandpa and his cousins kept beehives here.

The seaside cow pasture at the Grimes Ranch.
(Photo by Jenn Jackson)

"A bee."

"Yeah? I'd like to see that."

Grandpa suggested that perhaps he should come see it along with all the other dads. And so it was settled. Grandpa would be my fill-in for Take Your Dad To School Night. But I wasn't so sure this was a good idea. When I pictured the other dads, I saw men in business suits with briefcases and office jobs. Then I saw Grandpa standing next to them, with his hair all messed up and black dirt under his fingernails and no business cards. I hoped he'd at least remember to comb the food out of his mustache.

When the day finally came, I had convinced myself bringing Grandpa was a terrible idea. He would be so much older than the other dads that he'd call even more attention to the fact that I was without a real dad. All I wanted to do was blend in, and since starting school, I'd somehow done everything possible to make myself stick out. Now I was going to show up to Dad Night with an impostor, drawing even more puzzled looks. I wished that I'd just stayed home after all, and tried to think of ways to cancel our outing at the last minute as I waited in the living room for Grandpa to get ready.

Finally, he came out of the bedroom, adjusting his favorite bolo tie around his neck, the one with a nugget of turquoise mounted on a shiny silver square. He wore it only to square dances, funerals and weddings. I noticed that his jeans were creased down the front—he must have taken an unworn Christmas pair out of the cedar chest. His mustard-colored Western shirt had ivory snaps and thin gold metallic stripes. His hair was combed down, his stubble was gone and he smelled of aftershave. I checked his fingernails: clean.

We walked down the street to my elementary school. In one hand he held mine, and in the other he held a jar of honey for my teacher.

Inside the classroom, I guided Grandpa to the table of art projects and pointed out my bee. It was about the size of a loaf of bread, and I'd put effort into shaping it correctly, with six legs and four wings. I had unfolded two paper clips and poked them into the hardened newspaper for antennae. Grandpa picked the bee up and turned it to look at it from all sides, whistling appreciatively. Just then my teacher walked over and introduced herself, and he gently placed it back down.

"Quite a bee," she said.

Grandpa said he was happy to meet her, and held out the honey jar. She placed a hand over her heart as she reached for the gift.

"This is from your bees?"

"Yes, ma'am," Grandpa said.

"Incredible," she whispered.

I'd never heard Grandpa say *ma'am* before, and I giggled. He threw me a look that said not to blow his cover. He was on his best behavior, and so far, so good. No one had asked him who he was or why he was with me. We were a pair, and that's all that mattered. We stood close together while the other fathers talked to the class about their careers, and as I listened to stories about working in banks and courtrooms and on golf courses, I wondered what Grandpa was going to say. He didn't have a real job—one with a workplace and a boss and a paycheck. He just fixed things and kept bees. I worried he wouldn't have much to offer, or he might get flustered having to speak before a group. He once told me the great thing about being a beekeeper is that you can do it all alone, without having to speak to anyone. Grandpa was the kind of person who preferred to keep to himself, and he always used the minimal amount of words to convey a thought. I wasn't sure he was up to this.

The teacher called his name, and I let go of his leg. He walked to the front of his class and cleared his throat.

"I'm Frank, and I'm here with my granddaughter, Meredith," he said. "My family goes back four generations on the Big Sur coast."

I heard an interested murmur from the group.

Grandpa said his great-grandfather was one of Big Sur's earliest pioneers, William Post, who was eighteen when he left Connecticut in 1848 to become a whaler, finding work burning blubber into lamp oil and harvesting whale bones for corsets at the Monterey Whaling Station. Two years later, he married a local Native American woman from the Ohlone tribe named Anselma Onesimo inside the Carmel Mission. They built one of the first homesteads in Big Sur, the 640-acre Post Ranch, where they raised cattle and hogs and planted an apple orchard. They led cattle drives to Monterey, and packing trips to take hunters and fishermen into the Big Sur backcountry. And they had beehives.

Grandpa explained that he began keeping bees as a teenager, after a swarm landed in his yard and his father showed him how to capture it and put it in a hive. The bees quickly multiplied, outgrew their hive and started developing new queens—a sign the crowded colony was getting ready to divide itself by swarming. So his father showed him how to move the incubating queens and some of the bees into an empty hive to make a second colony. Within two years, father and son had five beehives behind their home in Pacific Grove, a small seaside community an hour north of Big Sur made up of Victorians squeezed close together on tiny lots.

The neighbors were patient, and somewhat fascinated, with his bees, he said. They became even more supportive when Grandpa placed a hive on the porch of a Japanese family that had been forced into a labor camp during World War II.

"No looters dared go near that house," he said.

While many were enthralled with his bees, his mother was losing her patience. After getting stung one too many times while hanging the laundry, she finally put her foot down and demanded he find more open space for his new hobby.

Friends and relatives in Big Sur were happy to help, and Grandpa relocated his hives to several coastal ranches where the bees wouldn't bother anybody. He placed beehives in a remote clearing at the foot of Garrapata Canyon, on a cousin's cattle ranch in Palo Colorado Canyon and in the nuns' walled-in vegetable garden at the Carmelite Monastery. People started calling him The Beekeeper of Big Sur.

Grandpa told tales of battling the seas to haul in nets of sardines along Cannery Row; he even made plumbing jobs seem exciting. He sounded like a superhero when he described tying himself to trees and dangling over the Santa Lucia cliffs to hammer rebar and secure pipes that directed water from a natural spring to Nepenthe, the landmark bohemian restaurant perched eight hundred feet above the sea.

I looked around the room. The kids were never this quiet in class.

"Sounds like you walked right out of a Steinbeck novel," one of the fathers called out, putting Grandpa on par with Monterey Bay's literary son. But Grandpa actually remembered Steinbeck, as well as some of the real-life marine biologists, hoboes and shopkeepers who appeared in his novel *Cannery Row.*

So Grandpa took the man's comment literally. "Steinbeck was a nice enough guy. Kinda kept to himself though. Doc Ricketts was more fun," Grandpa said. "He used to pay us to bring him frogs from the Carmel River for his lab. He had good jazz parties, too."

"Did you know Henry Miller?" another parent asked.

"Played Ping-Pong with him at Nepenthe once," Grandpa said. "He cussed a lot."

The kids pelted him with questions about bees. Did they sting him? How did he get the honey out of the hive? How does he catch a swarm? Grandpa started having fun with his audience. He said bees stung him all the time, but that meant he'd never get arthritis. He said he got honey out of the hive "very carefully," and told the students that he caught swarms with his bare hands. They couldn't tell if he was joking or telling the truth and stared at him, dumbfounded. Grandpa held the floor until my teacher politely interrupted so another father could have a turn.

He returned to my side, and I squeezed his hand. He had single-handedly erased all my prior social faux pas and given me a do-over at school. He was cool, and he told the class that I helped him with his bees, and therefore I could be cool by association. I should never have doubted him, and I felt sheepish that I'd thought he would mess it up. Grandpa was different, but that made him *better*, not worse. It made no difference anymore that he wasn't my dad. We were together now, and that's all that mattered. Grandpa squeezed my hand back.

"Very good," I whispered.

When it was time to go, I could feel eyes on us as we crossed the playground toward home. Grandpa had a stack of new orders for honey, names and phone numbers scribbled on punch napkins. And I had something worth more than money—his loyalty. Grandpa had claimed me before the whole class, and that meant he had been listening to me when I told him the troubles I was having at school. He had been thinking on it, and came up with his way of turning things around for me.

I learned on that day that Grandpa would stick up for me, in the same way a bee will defend its hive or die trying. He was letting me know, in his silent way, that he had made a promise. He would never leave me.

7

Fake Grandpa

1975—Winter

Shortly before Christmas, the Volvo, which had never fully recovered from its cross-country voyage, was gone. In its place was the strangest car I'd ever seen. It was Porta Potty blue and shaped like an avocado, wide in the rear with a long snout in front. It squatted low to the ground, and looked like its back end had been whacked off with a giant ax. A white racing stripe on each side swooshed all the way from tail to tip, tapering to a point just behind the headlights. It was an AMC Gremlin, advertised as the most affordable economy car in America, what Granny could buy for Mom on an installment plan.

Matthew and I approached cautiously, peering through the hatchback—one solid piece of glass on hinges. We saw white upholstery perforated with small dots, and plush ice-blue carpet, not yet stepped on. There was a radio with push buttons, and a white steering wheel as big around as a garbage can lid. The car gleamed with untouched possibility.

"I don't want you kids touching it," Mom said.

She was propped up in bed with all the pillows behind her back, reading the driver's manual.

"Can we go for a ride?" I asked.

She slapped the manual onto her lap. "What did I just say?"

"No touching," Matthew answered.

"That's right. Now go," she said, waving us away with her hand.

Granny disguised her gift to Mom as a loan, proposing a repayment plan once Mom found a job. It was a bribe of sorts, one that worked in Mom's favor. She didn't drive the Gremlin to a place of employment, but she took it other places. She fetched groceries now and then, and hit up a garage sale on the weekends. Ultimately, she never paid Granny back the two thousand dollars, but it was the carrot that lured Mom out of bed. The car gave her a modicum of autonomy, and I was grateful for this welcome blip of progress as Mom tentatively stepped back into society.

Then one day Mom broke her no kids in the car rule, offering to take us to Carmel. Mom opened the car door, and a clean and faintly chemical smell wafted out. She lifted a lever and the passenger-side seat bent forward, giving us a few inches to wedge ourselves into the back seat. This was a decade before children's car seats were required, and if there were seat belts in the car, they must have been buried in the cushions because we didn't use them.

"Watch where you're stepping!" Mom said, licking her

thumb and wiping away invisible scuff marks where our shoes had touched the seats. She bent down and shucked the shoes off our feet, knocked them together to dislodge the dirt and carefully placed them on the floorboards. I watched her walk around the car, get in and sling her purse into the passenger seat. The white interior glowed around us like an orb. Once seated, she revved the engine twice, then eased up on the clutch. She pressed too hard on the gas, and the car pitched forward and stalled.

"Shit!"

She checked us in the rearview mirror.

"Don't tell Granny I said that."

Mom turned the ignition and the car bucked again, harder this time. Matthew grabbed the seat in front of him for support. He shot me a mischievous smile. *Shit*, he mouthed. I turned away so he wouldn't see me laughing. There was something so funny about a foul mouth on a cute kid. Mom sighed heavily and sat for a moment with both hands gripping the steering wheel, her elbows locked.

"Are we going to Macy's?" Matthew asked. Santa was there inside the department store in Monterey, receiving gift requests.

"No, we're going to see your grandfather," Mom said.

A wrinkle appeared between my eyebrows. That made no sense. We already had a grandpa, the one who just yesterday hauled a Christmas tree from the back of his truck into the living room and covered it in blinking lights. I explained this to Mom.

"Quiet. I can't hear the clutch," she said.

Finally the engine caught, and she cautiously eased the

car down the driveway. Once she maneuvered onto Carmel Valley Road, she shifted into second, keeping it there even though the engine screamed for a higher gear. A line of cars backed up behind us on the two-lane road, and the driver behind us flicked his high beams through the large glass hatchback, lighting up the inside of the Gremlin like a pop of lightning. We ducked instinctively, but Mom ignored the tailgater as she pushed in the cigarette lighter and knocked a pack of cigarettes on the steering wheel until one poked out the top. She pulled it out with her lips and tossed the pack back into her purse. When the lighter popped out, she held the red coil to the tip and lit it.

"Grandpa isn't your real grandpa," she said, lowering the window and exhaling smoke out the crack. "My dad is. We're going to see your real grandpa. Granny's first husband."

A gong went off. The news that Grandpa wasn't my grandpa was preposterous. I'd never seen this other man, nor heard of anyone else claiming to be my grandfather. I dug my fingernail into the white seat, trying to poke a hole. She was trying to tell me that Grandpa wasn't good enough somehow, but I refused to believe it. I fumed in the back seat, at Mom for so casually dismissing Grandpa, and at this stranger for taking Grandpa's place without my approval. Mom held her cigarette out the window to let the wind suck off the ash, then brought it back to her lips.

"Grandpa is my grandpa," I insisted.

"No, he's your *step*-grandfather."

My mood was good and foul when she eventually turned off Highway 1 onto Ocean Avenue, and we descended a steep hill that bottomed out before Carmel's downtown shops. She steered the growling Gremlin away from the main street and up through a wooded neighborhood of cheery cottages that looked like gingerbread houses with icing. The roofs were thatched and wavy, some decorated with flags or weather vanes. Windows had flower boxes, doors were flanked by lanterns. Everywhere I looked I saw cobblestone walkways, and the homes had names instead of numbers: *Country Comfort*, *A Whistle Away*, *Sea Shadows*.

These were the homes that originally belonged to painters and poets and actors when Carmel began as a seaside artists' colony at the start of the twentieth century, but were now heavily renovated and occupied by descendants or wealthy outsiders. Each was unique, yet all were the same in that they were the kind of homes I'd never been in before. I felt suddenly self-conscious, and worried about just what, exactly, Mom was getting us into.

My foul mood got a little fouler. Mom took the car slow around an oak tree that was growing right in the middle of the narrow road. In the twisting backstreets of Carmel, trees sprouted here and there from the asphalt with reflective tape around them, directing drivers to respect the landscape and please go around. Whoever had built the roads hadn't had the heart to chop them down, and locals were accustomed to driving slowly to go around them.

Mom pulled into a parking space above a house that clung to a hillside overlooking a canyon of Monterey pines. We walked on a narrow balcony that wrapped around the house, until we were facing a big red door flanked by two Chinese lion sculptures, one with a paw resting on a ball and the other with its paw on a cub.

Mom smoothed her skirt and stood up taller, then knocked. As if someone was standing on the other side peering through the peephole, the door whipped open immediately, and a short, thin man in pressed khakis, tasseled loafers and an oxford shirt stared at us. His white hair was precision-combed as if he were still in the military. He had a pink, expressionless face, dark eyes and a mouth that relaxed into a natural frown. I had never met him, yet I felt he was already disappointed with me. He and Mom looked at each other in silence. I had a sudden urge to bolt for the car.

"Sally."

"Dad."

He opened the door wider and motioned us in. I reached for Matthew's hand.

Our steps echoed as we entered what appeared to be a modern art gallery instead of a home. The two-story architectural home was cold and impersonal, empty in the center with a top floor that formed a ring around the one below. From anywhere on the upper balcony, you could peer over the ledge down to the first story, which had a decorative concrete floor with slices from a massive redwood tree trunk embedded in it. The wall facing the canyon was all glass. A floating staircase connected the two levels. The walls were decorated with Chinese paintings of fog-filled mountaintops and fighting warriors. A towering silvertip Christmas tree as giant as the one at Macy's rose from the bottom floor. There wasn't a speck of dust anywhere in this showpiece of a house.

Mom instructed us to say hello to our grandfather. I gave a wan smile. He shook my hand and contemplated me. I had the unnerving sense that I had to apologize for something. My heartbeat sped up and I swallowed apprehensively, waiting for him to tell me what I had done wrong and what my punishment would be.

I heard footsteps behind me, and his wife broke the spell, swooping up to greet us in flowing robes, with a necklace of bulky red rocks and jade rings on her fingers. She had salt-and-pepper hair, a square jaw and prominent cheekbones, and stood a foot taller than her husband. She explained that she was going to make us a wonderful tea. She said the name of the place it came from, and we all stared at her blankly, and then she explained it was a very high mountain in China and the tea was served in the temples they had visited. We were led to a sitting room on the top floor, next to the kitchen, and took our places in rigid, antique Chinese chairs. We sat with Mom on one side of the room, and her father sat opposite us. Mom's eyes darted around at the tapestries, out the window, anywhere but at her father. She spilled sugar stirring it into her tea. She hated tea and only drank coffee at home.

I was afraid to touch anything. Matthew remained angelically still in his chair as his eyes scanned this strange place. There wasn't a toy anywhere.

I could tell that Mom was already regretting this visit. It was obvious that she and her father didn't like being in the same room, that they had no idea what to say to one another. The air crackled with unspoken resentments.

I would later learn that he'd been a brutal father, teasing her mercilessly for being overweight and bruising her for the slightest transgressions such as not cleaning the house to his satisfaction or scowling at him. His unpredictable tirades consumed her childhood, and Granny's happiness, as well. They lived in fear of him until finally, when Mom was nineteen, her parents divorced. Mom was euphoric when he left, relieved to never have to speak to him again. Now here she was, a dozen years later, in his tearoom. It could have been that Granny had forced her to go, to ask her estranged father for financial help. But I think it was more likely she was drawn by mixed emotions of curiosity, hope and need. Using the holidays as a pretense for reparations, she was testing her father to see if he had changed, if he felt remorse and if he would help her get back on her feet.

He cleared his throat.

"Well, Sally, how are you getting along?"

She said things were okay, but that they were hard, too. She told him she might be able to get a job as a bank teller or a nurse's aide at the hospital.

"That's good, Sally. But what about doing something with your sociology degree?"

Mom began chipping at her nail polish.

"Have you thought of graduate school?" he pressed.

Mom said she couldn't afford grad school. She had two kids to feed.

I could see a light behind his eyes turn off.

"I'll be fine, Dad."

I wished I could say something to change the subject, but I felt frozen in front of this sudden grandfather. I tried to picture this man, his cashmere sweater draped over his back and the sleeves tied together in front, living among his art books and carved dragon sculptures inside our small, single-story country house, and it made no sense. He didn't look like the kind of person who chopped wood to build a fire, or weed-whacked outdoors, or came in contact with dirt, ever. In his house, everything was on display and seemed rarely used. In ours, every corner was jammed with worn-out things. My grandparents collected rubber bands into balls, smoothed and reused tinfoil, and saved every paper bag. We were from different tribes, and I couldn't picture this man in ours. It seemed as if Grandpa's piles of junk had sprouted from the ground and laid claim to the property centuries ago.

The wife appeared with a tray of butter cookies, placing them on the coffee table while telling us about their recent trip to China. The conversation blurred into a boring adult drone as they detailed treks to ancient sites dating back to such-and-such dynasties, and I wondered if I could fall asleep with my eyes open. I lifted my teacup and saw what looked like a piece of seaweed floating in it. I set it back down on the lacquered coffee table. Mom fake-listened to his stories but her mind was elsewhere, her gaze fixed on a point on the wall just above his shoulder. I watched his mouth move without listening to the words. When he finished talking, another long silence descended on our group. Her father cleared his throat again.

"Want to see the rest of the house?" he said.

He took us on a loop of the top floor, showing us the kitchen and the bedrooms with screens that slid into different wall positions. There was a small library with swords mounted on the walls, and then we descended to the bottom to the big open room with the Christmas tree. Downstairs was an office, several more rooms with sliding walls, and a piano. New Grandpa turned his attention to me, asking me whether I liked elementary school. I told him it was fine. Then he asked me what I wanted to be when I grew up. Nobody had ever asked me that.

"I don't know."

"Well, it will be a doctor or lawyer, one or the other, right?" he said, pinching my cheek. It hurt, and I took a step back, rubbing my face. Mom's face was starting to flush in anger.

"She's going to be whatever she wants to be, Dad," she said firmly.

Again, there was nothing more to say. Mom looked out toward some storm clouds gathering over the canyon and frowned. Her father led us toward the Christmas tree, which was encircled by a throng of presents. He reached for one and handed it to Mom. She opened a furry V-neck sweater from Neiman-Marcus in a shade somewhere between green and brown, like pond water. Mom did not wear sweaters. She said it was lovely and put the box on the floor.

"For the young man," he said, handing Matthew a gift. When my brother unwrapped a Tonka dump truck, he immediately tore open the packaging and started pushing the truck around on the floor while Mom's father kept a watchful eye on his vases.

My gift was a ceramic speckled goose egg to hold jewelry. I didn't have anything to put inside it, but I thought it was beautiful and delicate, something a nice lady would have in her house. It made me feel grown-up to be entrusted with a breakable. My mood shifted slightly.

We stayed for a few more cookies, and then Mom stood up and said it was time to go. Our host didn't try to argue. He thanked us for coming and walked us back toward the big red door. He didn't hug anybody goodbye; he just stood with one hand on the door and waved with the other.

Mom speed-walked to the car. She slammed the door, jammed the key into the ignition and screeched out of the parking space in reverse. She was so mad, she forgot to make us remove our shoes. She jerked the steering wheel to negotiate the curvy Carmel roads, and Matthew leaned toward me and whispered, "Jell-O." I nodded, and we let our bodies go slack in the back seat, swaying together as Mom swung the car side to side. Mom was muttering under her breath, and hitting her thigh with her fist. Then she started talking to no one in particular.

"Did you get a load of that house? I could use a little help, too, you know. But *noooooooooooooo*!"

She was shaking, and she might have been crying, but I wasn't sure. Matthew and I flopped to the left, then the right, then back again, absorbed in the task of turning our bodies to gelatin.

"I don't know why I try. Stupid, stupid, STUPID! I shouldn't even give him the time of day. He never gave a shit about me, that's for damn sure!"

Matthew was about to practice the *S*-word again, and I quickly put my hand over his mouth. Mom kept on talking to her cigarette. She banged on the steering wheel between sentences.

"After everything he did to me!"

Bang!

"People never change."

Bang!

"Same old bastard!"

Bang!

Matthew and I kept our bodies pressed close even after the road evened out. We braced ourselves against the torrent of angry words coming our way, keeping each other comforted as much as anyone can when trapped inside a closet with a person using a megaphone. Mom was hollering now, her words ricocheting inside the car and bouncing off the walls, colliding and smashing to bits over our heads. All the things she wished she'd said to her father came pouring out in our rolling confessional. She didn't need him. He meant nothing to her. She wished him dead. She would never waste her breath on him ever again.

I wanted to comfort her, but Mom seemed unreachable to me, lost in her own memories that were too awful to share. Whatever had happened to her was too big for me to fix with words.

I wanted to hurry home so she could get back in bed where it was safe. Just catching this glimpse into her past, I became more sympathetic, and promised myself not to get so angry with her for staying in bed. The world was hard for her, and I needed to be patient because there were deep reasons from her past that made her give up on the present.

As she turned under the canopy of walnut trees leading into our backyard, she shook her index finger and proclaimed, "I tell you what… That is the *last* time he will ever see me, or you two kids again!"

The knot in my stomach released. The double grandfather problem vanished, nothing but a magician's trick gone up in smoke. When I woke up on this oddest of days, I had one grandpa. By midday I had two. Now I had one again. I guess it would have been baffling for most kids to gain and lose a grandfather in the space of twelve hours, but to me it was just another example of how relationships in my family shifted with the sudden winds. One day a person was in, the next day they were forgotten history. I was starting to get used to the impermanence of people, of places, of promises. Everything changed with Mom's fluctuating moods, so it was better to let her words slide on by without assigning them too much meaning. It didn't matter anymore because that impostor grandfather had become an unmentionable. He was never real to me, anyway. But I was still keeping the pretty ceramic egg.

Mom continued cursing under her breath as she entered through the front door. Granny was stretched out on the rug with her daily libation, and when Mom passed by without so much as a hello, Granny looked up with a bemused smile, and swirled the ice cubes in her plastic cup.

"So, how is good-ole-what's-his-face-my-first-husband?" Granny called after her.

The bedroom door slammed in response.

"Told you so," she said, shrugging her shoulders at Matthew and me. My brother placed his new toy before her.

"Lookit my truck."

She picked it up and inspected it from all angles.

"That is a *very* fine dump truck. Why don't you go outside and fill it with dirt?"

Matthew didn't need to be told twice. He clutched his prize and whizzed outside to the sandbox, and I followed, seeing as I didn't have anything better to do. While Matthew pushed the truck and made engine noises, I picked the fallen toyon berries out of the sand and lined them up to make a roadway for him. Our sandbox was a simple square made from four redwood boards and just big enough to fit both of us, filled with sand Grandpa had liberated from Carmel Beach. The sand was so white and clean that it squeaked when we squeezed it. Matthew was packing and dumping truckloads of sand to make buildings when we heard the rattle of Grandpa's truck and the pop of walnuts crushing beneath his tires.

"Grandpa's home!" Matthew chimed.

Grandpa parked under the carport, set his lunch box and keys on the hood, and Rita ran toward us and hopped into the sandbox to dig. Grandpa plucked a yellow bloom off a mustard plant and chewed it as he walked toward us.

"Whatcha got there?" he said, reaching for the truck.

He gave it a few test pushes in the sand. "It's good," he said. "Strong engine. Where'd you get it?"

I told Grandpa we went to Carmel to meet Mom's dad. Grandpa nodded silently and sat down on the edge of the sandbox, waiting for me to continue.

"Mom said you're not our real grandpa."

Grandpa was quiet for a moment, thinking. Then he lifted me up and sat me on one knee. He reached for Matthew and put him on the opposite knee.

"Now you two listen, and you listen good," he said. "Pinch my arm."

We checked his face to see if he was serious.

"I mean it. Hard as you can."

I squeezed and dug half-moons into his forearm with my fingernails.

"Do you feel skin?"

We nodded.

"Then I'm real. I'm your grandpa."

Satisfied, Matthew hopped off Grandpa's knee and ambled back into the house. I felt better, but something was still gnawing at me.

"What's *step* mean?" I asked.

"*Step* just means you're lucky because you get to have more than one grandpa."

"But Mom said..."

Grandpa leaned in until our noses were almost touching and locked eyes with me. "Sometimes she gets confused," he whispered softly, so only I could hear.

He was telling me that it was okay to make up my own mind about whom I wanted for a grandpa. And it was an easy choice because Grandpa's life had room for us, and wasn't complicated by tangled family history. He was the adult who looked forward to seeing us, enjoyed teaching us new things and truly cared about our opinions. He loved us the way a parent should.

A shadow crossed over the sandbox and Grandpa looked up at the purple clouds threatening rain. "I need to check a hive real quick. Wanna put on your veil?"

I followed him to the back fence and stood a few feet away as Grandpa disassembled the hive. First, he set the lid upside down on the ground, then he wedged his hive tool underneath the top box and cracked the sticky seal the bees had made. He twisted the box loose, and, cheeks puffing out with effort, set it on the upturned hive cover so the bees on the bottom wouldn't get smashed. The uppermost boxes on the hive were where the bees stored their honey, and they weighed up to fifty pounds when full. This hive had two supers, and Grandpa took them both off without examining them. He could tell by the weight that they weren't yet full.

Also, this time of year he wasn't interested in taking honey from the bees. They'd need it over winter for their own food. He removed honey during the nectar flow in spring and summer, and took only the surplus so they'd have enough to eat. Today Grandpa's aim was to get down into the bigger brood boxes at the base of the hive, into the nursery where the queen lays her eggs in the sheets of waxen honeycomb.

This particular hive had been giving him trouble all year. In spring, half the colony had swarmed with the queen, and the workers that remained behind produced a second queen. Soon after, that one absconded as well. While it's natural for a hive to propagate this way, each exodus was a setback for the colony, forcing them to expend energy and time rearing a new queen and waiting for her to get mated and start laying eggs again.

Today Grandpa was hoping to find eggs in the nursery, the telltale sign that the queen was healthy and the colony had righted itself once more.

The guard bees were fussy and encircled his head as he worked, every once in a while one would break patrol and head-butt him to alert him to the colony's diminishing patience. They weren't ready to sting just yet, but they would if this inspection took too long. It was the afternoon, and the bees were returning from their foraging trips to tuck in for the night, having enjoyed the best hours of the day with full sun. They didn't appreciate the cool afternoon air and sunlight invading their home at the precise moment they were winding down for the day, huddling together inside for warmth.

Once Grandpa had exposed the box containing the nursery, he lifted the outermost wooden honeycomb frame from the row of ten inside, examined both sides of the honeycomb and quickly determined it was filled with honey. He set it down and propped it against the fence. The next frame he pulled out was similarly filled with honey. The third frame had empty comb in the center, with a smattering of stored pollen and a little honey stored near the top. When he reached the frames in the middle of the box, he removed one and found it covered with nurse bees scurrying over the honeycomb, darting their heads into the hexagons. He brushed a few aside with his finger, and tilted the frame back and forth in the dying light so he could see if the bees were feeding the larvae inside the cells.

"We're in business!" he exclaimed. He held out the frame so I could see little white larvae curled into C's in the bottom of the cavities. Those small worms were four days old. He pointed at another area of the honeycomb, and I saw vertical white pins, the fresh eggs. The nurse bees were so intent on feeding that they stayed on the frame, paying no attention to us as we turned the frame side to side to examine it.

"Is the queen here?" I asked.

"Not on this one," he said. "Need to keep looking."

Just then I felt a raindrop on my arm.

The rain picked up quickly, drops pattering on the frame in Grandpa's hands. Now the nurse bees were lifting their heads and taking in their surroundings, running to one another and tapping antennae in a mad frenzy. They were clearly perturbed by the strange sight of water in their nursery.

"Better close up," Grandpa said. He took a few steps toward the hive, then stopped in his tracks and stared at the frame in his hands. "Well, I'll be darned!"

He wheeled around and held the frame aloft. Where just seconds before the nurse bees were stumbling in all directions and knocking into one another, agitated by the sudden rain, now several hundred were lined up in perfect rows like corn on the cob. They were organized as precisely as a battalion, all facing the same direction with their heads north and their wings interlocked, forming a tarp over their precious eggs. They stood together motionless, their posture rigid, their wings tight together like Spanish roof tiles, protecting the next generation from the rain.

Grandpa had successfully convinced me that bees were smart. But I didn't know bees were capable of love. I marveled at their sacrifice as they took the brunt of the raindrops on their backs and diverted the water away from the young into little rivulets formed by their overlapping wings. How long would they stay like that if we didn't put the frame back into the hive? They looked so determined that I imagined the bees would stand guard until the rain passed, or until they got so waterlogged or cold that their hearts stopped beating.

It defied logic that nurse bees would know how to do this. Nurse bees remained indoors—either in man-made hives or in hollowed-out trees or inside the walls of houses, wherever colonies set up a dry home. They were "house bees" that didn't venture outside to forage, not until they had learned to fly long distances and had matured into field bees. So they weren't familiar with rain. How could they suddenly know how to align themselves into a makeshift umbrella? And how did they send out the signal so quickly, so that they simultaneously snapped into formation?

I stood there, gawping.

"Isn't that something," Grandpa said. "A friend of mine said he saw this once, but I didn't believe him."

"How'd they do that?"

"You'll have to ask Mother Nature," he said.

Grandpa slid the frame back into the safety of the nursery, restacked the hive boxes and placed the cover on top, securing it with a brick. The nurse bees would soon dry out in the warmth of the hive.

As we walked back to the house for dinner, I thought about what I had just witnessed. I had seen insects displaying unconditional love. The nurse bees huddling against the rain were not the parents of the babies they were protecting. The queen was. But they put themselves in harm's way because nurse bees were hardwired to raise the queen's offspring. They were surrogate parents, just like Grandpa was for my brother and me.

It took thousands of nurse bees to care for so many eggs, so a bee colony divided up the responsibility. It didn't matter that the nurses were unable to give birth; they still knew what to do. Each bee had equal love, and there was no distinction inside a beehive between "step" and "real."

The bees had just confirmed who my real grandpa was.

8

First Harvest

1976—Summer

Most months of the year, the honey bus remained dormant. But after the spring nectar flow, toward the start of summer, Grandpa began keeping an eye on the thermometer nailed to the garden fence. When the red line rose above ninety, the conditions were ideal for extracting honey. Heat made the honey runny, so it could pump faster through the pipes inside the bus. If there had been exceptional spring rains to produce abundant flowers, he could increase his output and bottle nearly one thousand gallons of honey.

I had been asking Grandpa all spring if I could help him with the harvest. He didn't let me into the honey bus last year because he said I needed to be bigger. Now that I was six, and I had gone up two shoe sizes, I was campaigning hard for admission. Each morning I checked the weather to let him know I was monitoring the situation and ready to report to duty, in case an ideal harvest day arose.

And finally it happened. One July morning I awoke to the chorus of the cicadas, shrieking in the heat. I got out of bed and pulled back the curtain and saw Rita curled in the shade of the apricot tree, panting. This hot this early could mean only one thing—the high holy days of harvest were upon us. I hustled outside in my pajamas and checked the thermometer. Almost ninety already. I found Grandpa bent over a tall stack of pancakes at the dining room table, and gave him the good news.

"It's honey weather," he pronounced.

Grandpa slowly chewed a mouthful like he was considering a complex algebra equation, and then took a long slurp of his coffee in that unhurried way that old people have about everything. He folded his paper napkin in half, then in fourths, then daintily dabbed the corners of his mustache before he cleared his throat. I held my breath, waiting for his verdict.

"You'd better put on overalls," he said.

He went back to stabbing his pancakes as if the earth hadn't just tilted off its axis. I tore out of my pajamas and changed into overalls in record time. I didn't know why Grandpa had changed his mind and decided to let me in the honey bus, but I wasn't about to ask any questions lest he reconsidered.

Before the honey bus was the honey bus, it was used by the U.S. Army to transport soldiers from the Fort Ord military base, just north of Monterey, to other outposts along the California coast. Ford Motor Company built it in 1951 as part of its F-Series, its first postwar truck and bus redesign, and sent the twenty-nine-passenger bus to Fort Ord to fulfill a government purchasing order made during World War II. As new equipment kept arriving despite the end of the war, the base became overcrowded with a glut of vehicles and began selling off some of its barely used inventory. A friend of Grandpa's in Big Sur bought the bus at auction to cannibalize its six-cylinder engine for his own truck. He put a lighter duty engine in the bus, and sold it to Grandpa in 1963 for six hundred dollars.

Grandpa was inspired to build a portable honey house after reading a story in his beekeeping magazine about beekeepers who installed honey spinners on the flatbeds of their Ford Model A trucks, so they could drive up to their apiaries and harvest right on the spot. But Grandpa thought that was silly because if you harvest outdoors, the bees will find the honey and go into a robbing frenzy over it. With a bus, he could drive to his bee yards and extract honey in a closed environment, without getting stung. He ripped out the bus seats and gave them to friends, who installed them in the backs of their pickups, and built his honey factory inside with parts from his sprawling collection of spare junk.

He was mighty pleased with himself, until he tried driving the one-and-a-half-ton honey bus into the steep Big Sur canyons and nearly got it stuck on the switchback dirt roads more than a few times. After that, he steered clear of his more remote apiaries and drove the bus strictly to the bee yards he kept close to the highway.

Also, he didn't figure how expensive it would be to keep a bus running. His F-5 sucked down gas, and he had to fork over several hundred a year just for insurance and registration. So, to Granny's utter horror, he parked the green monstrosity behind the house in 1965, removed the engine and gave it to a friend. Back then, Carmel Valley was still a country place where real cowboys hunted wild boar and scooped crawdads out of the river, before the tourists started demanding espresso at the Wagon Wheel breakfast counter, and according to Granny, stinking up the place with their cologne and talk of race cars and golf swings. It was a time when people could leave sputtering buses in their backyard, and no one blinked twice.

I followed as Grandpa cut a path through waist-high foxtails to the bus. His dusty Levi's kept sliding down his butt, and he didn't bother with a shirt, revealing a barrel chest that was a shade somewhere between cinnamon and rust. His sinewy arms ended in two bear paws that were covered in cracks, pockmarks and scars from work. The top half-inch of Grandpa's left pointer finger was missing, and his nail had grown all the way around it, like a helmet. An accident during high school shop class, he said, when he was cutting metal to make air raid sirens for the war. We weaved around small middens of pipe fittings and broken pottery, and stopped before an old wooden highway sign propped against the back of the bus: *Pfeiffer State Park: 5.1 miles*, with an arrow below it and the words *Lunch This Way*.

My anticipation swirled as he climbed atop the staircase of wooden pallets at the back door, and felt around for the piece of rebar he kept out of my reach on the roof. He inserted one end of the bar into the hole where the door handle used to be, twisted and popped the lock. The door opened with a soft sucking sound, and he lifted me up and put me inside the bus. He followed and quickly slammed the door to block the handful of honeybees on our tail. They were attracted to the honeycomb that Grandpa had stacked inside the honey bus, giving off an aroma of vanilla, butter and fresh dirt that I immediately recognized as the scent of Grandpa's skin. It was like the air inside the honey bus had its own flavor.

Inside I saw white hive boxes stacked in towers along the wall opposite the machinery, reaching almost all the way to the roof. I started counting and reached thirty-seven and stopped. We were going to make buckets and buckets of honey, I figured. Grandpa took the lid off the nearest hive box, and pulled out one of the wooden frames of honey and admired the delicate hexagon cells sealed with a thin layer of yellow wax. He held it up to the light, letting the sun illuminate the amber nectar like a stained-glass window. He let out a long, low whistle of satisfaction.

"That's a goodie," he said, handing it to me so I could feel its weight. It felt like a heavy dictionary, easily three pounds of honey.

Grandpa took it from me and eased it back into the box with the nine others just like it. He walked down the narrow walkway toward the front of the bus, his footsteps making sticky sounds on the black rubber floor like he was walking on human flypaper.

"Does this thing work?" I said.

I reached for the rope pull, gray and frayed with time, and a bell clanged. Grandpa shot me a look from behind the driver's seat, where he was pouring gasoline into a lawn-mower engine that he'd rigged to power the honey spinner, and I let go of the rope. The motor whined and wheezed as he yanked a pull-cord to start it, but eventually it caught and steadied to a pounding rat-a-tat that vibrated beneath my feet. The whole bus shimmied. To keep exhaust fumes out of the bus, Grandpa had drilled a hole in the floor and fed a metal pipe from the lawn-mower engine to the outside.

"Now c'mere, let me show you something," Grandpa shouted over the din, waving me toward the spinner. I peered inside the waist-high metal tank at a flywheel with a rectangular cage dangling from each of its six spokes. The cages were just the right size to accommodate one frame of honey. When the flywheel spun, the honey flew out of the honeycomb and dripped down the insides of the extractor. The honey was then pumped up a pipe, and directed through a network of smaller pipes suspended from the ceiling handrails with fishing line. The honey poured out of the pipes into two storage tanks.

I gave the flywheel a push, but I didn't realize it had a lock switch. Grandpa gently moved my hand away.

"Rule number one. Don't touch stuff. And especially don't put your hands in the spinner. Unless you don't like your hands very much."

I glanced down at his shortened index finger and instinctively backed away from the spinner. I had to be careful not to get myself kicked out of the honey bus. I stood quietly, hands in pockets, so I wouldn't be tempted to touch anything else. As Grandpa prepared our workspace, moving jars and boxes out of our way and greasing the gears, I scanned the bus and to my delight discovered two grab bars running the length of the ceiling. This was excellent, my own monkey bars to practice on so I could do tricks on the playground with the other girls. Forgetting my vow of just a minute ago to be good, I hopped up and grabbed the two bars and swung back and forth, gaining momentum until I was able to get my legs up and over one of the bars and dangle upside down by my knees. Grandpa saw me and grabbed the opposite handrail, lifted his feet from the ground and hung opposite me.

"Oh yeah?" he said. Then he reached out and tickled my armpit, making me scream until I couldn't stand it any longer and swung myself back down to the floor.

"You ready to get to work now?" he said.

I followed him toward the back of the bus, to a long metal basin littered with wax curls and dead bees. He handed me a double-edged knife, its foot-long blade blackened with layers of burnt honey. It had a hollow wooden handle, and two rubber hoses inserted into it and secured in place with clamps. The hose trailed through a hole in the wall of the bus, outside to a copper pot of boiling water atop a propane burner.

"Careful, that's steam in that hose," Grandpa warned. "That's why they call it a hot knife. It'll burn you good."

I held the weapon in front of me with straight arms, like a knight with a saber, and waited for instructions. As the blade heated up, the crusted honey on it began to glisten and smell like caramel, and a curl of smoke wafted off the tip. I held it as far away from my body as possible, while Grandpa placed a honey frame on its short side on a nail protruding from a crossbar over the trough, and held it upright with one hand. He put his other hand over mine and guided the hot knife from the top to the bottom of the sealed honeycomb, holding the blade at a perfect angle to slice the wax seal, revealing the gleaming honey underneath. The wax curled away from the honeycomb and fell into the catch basin. It took a gentle touch to remove a thin layer of wax without carving into the honey.

"Now you try."

He let go of the handle and the knife became unwieldy in my small hands. I was afraid of it, and let it slip out of my hands into the trough, where it started smoking on the spilled honey. Grandpa fished it out, and used a wet rag to clean the honey off the handle. Maybe Grandpa had been right; I wasn't old enough to harvest honey.

"Use two hands."

It was getting so hot in the bus that my hands were sweaty, and I couldn't keep a good grip on the knife. I tried to steady the blade just like Grandpa had done, but I ended up poking it into the honeycomb and gouging out a big chunk of honey.

"Here," he said, reaching to guide my hands again. We uncapped several dozen frames together with his hands over mine, until I learned to feel the give of the wax and could slowly apply the right pressure to slice on my own. It took me a long time to uncap both sides of one frame, but Grandpa waited patiently, praising me and taking over when I got frustrated. Eventually I could remove a thin layer of wax and leave most of the honey inside the comb.

It was sweltering now, but we couldn't open the windows because there weren't any screens to keep the bees out.

Grandpa turned on a rotary fan near the driver's seat, which helped circulate the air, but added to the cacophony inside the bus. Then he stepped out of his jeans so that he was wearing only his tighty-whities and Chuck Taylor sneakers.

"Much better," Grandpa shouted over the ruckus. He reached into the uncapping trough, and pulled out a piece of sticky wax and popped it in his mouth.

"Chewing gum." He grinned.

He was always trying to convince me that the most disgusting things were downright delicious, like liver or blue cheese. He offered me a piece of honeycomb, and I tore off a tiny piece and tentatively bit down. It tasted like every candy I loved mixed together—I first sensed coconut, then red licorice and a blast of butterscotch. The texture was like a warm marshmallow melting on my tongue, and I couldn't believe I had had no awareness that a pleasure such as this existed. I chewed until the wax turned cold and then mimicked Grandpa, removing the wad from my mouth and tossing it back into the tub and grabbing a new warm piece. Grandpa took a few steps back and then winked at me. Then he spat his wax into the air like he was launching a watermelon seed, landing it in the basin. I took his cue and shot my wax into a big arc just as he had done.

"Two points!" he said, going all the way to the opposite end of the bus for the long shot. He spat and missed, the wax ball landing at my feet. He retrieved it, and as he stood back up, he leaned toward me as if he were going to tell me a secret.

"How's it going with your mother?"

I shrugged.

"Are you getting along?"

"I guess so," I said.

"It might take her a while to get better, you know," he said.

"Yeah."

Sealed away inside the bus, where he could speak his mind out of Granny's earshot, Grandpa's personality changed. He was talking to me as if I was his equal, and it took me a moment to adjust. I could sense he was trying to tell me something important, searching for the right words yet not wanting to upset me or tell me more than I could handle. He turned back to slicing wax, but kept talking to me in this new, grown-up way.

"She can't help the way she is."

His words hung in the air. What way was my mother, exactly? I knew that sadness followed her into every room. I knew she had to stay in bed because she had so many headaches, and that she really didn't like her father. By listening to my classmates, I had figured out by now that other moms went to work, came to school, cooked dinner. Mine slept through Christmas and left my brother and me personal checks, instead of actual gifts, under the tree. Our mother was different. But now Grandpa's words poked at me. Why was Mom "that way," and why couldn't she help it? What was wrong with my mother? Grandpa had admitted something to me, maybe something that I wasn't supposed to hear.

"She can't help what?"

Grandpa turned an empty hive box on its short side and sat on it like a stool. He wiped his brow on the back of his arm and faced me. I could tell he was choosing his words carefully.

"Your mother loves you."

I waited for him to continue. He tried again.

"Sometimes it's hard for her to show it."

"Why?"

Grandpa looked up toward a spider spinning a web in one of the oblong windows at the roofline. I could tell that I'd asked one of those questions for which there is no answer. In the silence that stretched between us, a heavy sadness pressed down on my chest, and suddenly I needed to sit down. I pulled an empty hive box near him and made my own stool.

"Have I told you about scout bees?" he asked.

I shook my head.

"Scout bees are house hunters. If their home is not right—too crowded, too damp—they go searching for a better one."

I wasn't sure why he was telling me this, so I waited for him to continue.

Scout bees are the risk-takers, the ones that convince a hive to swarm, he said. Days before the bees pour out of their hive in a massive cloud, scout bees investigate the neighborhood looking for a better place to live, exploring tree cavities, insides of chimneys, even the walls of houses. They wait for a nice, sunny day and then race through the hive, shivering their wing muscles against other bees to motivate them. Their excitement is infectious as the temperature rises inside the hive and all those flapping wings come together like a drumbeat. The bees get louder, and louder, and when they are at a roar, on some hidden cue the swarm pours forth from the hive entrance, whirling into a horde up to thirty feet across with the queen somewhere in the middle.

I imagined a firework of bees in the sky, tens of thousands of black dots swirling and then coming together as if through an invisible funnel.

"How do they decide where to go?"

"They dance."

By now I'd learned that Grandpa was never kidding when he talked about bees, no matter how unbelievable his stories seemed. He had me convinced that bees could do anything. I knew bees communicated by scent, sound and touch. So, why not movement, too? Now he was saying that foragers will dance inside the hive to tell the bees where to find nectar-rich flowers. The scout bees dance right on top of the clustered swarm to tell it where to relocate.

"The dance is like a map," Grandpa continued. "The dance steps tell the bees the address of their new home."

"Can I see?"

"See what?"

"The bees dancing."

"If you're lucky, we'll catch them doing it sometime."

Grandpa stood and began getting ready for the first spin. He reached into the uncapping trough for the honeycomb frames we'd unsealed with the hot knife, and slid them, dripping with honey, into the cages that dangled from the flywheel inside the extractor. Once he'd filled each cage, he unlocked the flywheel and paused before setting the spinner in motion.

"I don't want you to get too upset about your mother.

You're smart like a scout bee. One day you'll find your own way."

I decided right then and there that the scout bee was my favorite bee of all.

"Go ahead, flip the handle," he said, pointing at the lever near the lip of the spinner.

The flywheel spun and picked up speed, whining until the cages below blurred. The honey flew out in thick cords at first, thinning with each revolution until it became glistening gossamer threads, signaling it was time to crank the handle protruding from the top of the flywheel the other way and reverse the spin. It took a few minutes per side, give or take, depending on how full the honeycomb was.

Nearly a foot of honey collected in the basin, so thick and shiny that we could see our faces reflected in it. The pump kicked in and gulped at it, sending languid bubbles to the surface as it pushed the honey up through the pipes. The plumbing reverberated as the pump forced the honey up a main artery leading from the spinner's basin, all the way to the ceiling, where the tube branched off at a Y-joint into two smaller conduits. From here the honey was channeled past the passenger windows and toward two fifty-gallon storage tanks behind the driver's seat. The honey pipelines came to an end just above the openmouthed tanks, and were suspended in place by metal wire Grandpa had secured to the ceiling handrails with heavy-duty plumbers' tape. I kept vigil over those spouts, bewitched.

"Here it comes!" Grandpa said.

The first rivulets of honey bubbled out of the pipes, and cascaded into the holding tanks. It was pretty, like a girl's blond hair undulating in the wind. I remembered Grandpa told me once that one bee makes less than a thimbleful of honey in its whole life. So much was spilling out now that it must have taken a million-million bees to make it all.

We worked all day, until the sun started to sink behind the Santa Lucia Range, turning the mountains from dark green to gray, until we had nearly one hundred gallons of honey. I lost myself in the movements of lifting frames and slicing wax, and imagined we were worker bees inside our own hive. The whir of the extractor sounded like the hum of a colony, drowning out our voices so we had to communicate mostly by hand gestures. We nudged each other in this direction or that, shaking one another by the shoulder to convey something important. If we were at either ends of the bus, we had to wave and dance like a bee to get the other's attention.

Grandpa cut the motor with the last rays of the sun, and even after the bus shuddered to stillness, my ears continued to ring. My arms ached and my throat was dry. We had slight wax sheens on our hair and skin, and we smelled of butter and sage. I had never worked so hard that my body felt asleep before bedtime. Grandpa lifted the gate at the bottom of the holding tank, held an old mayonnaise jar under the spout, and filled it with honey. He reached for a roll of square white labels with red lettering and slapped one on the jar:

WILDFLOWER HONEY

U.S. Choice
From the Big Sur Apiaries
E. F. Peace

"Here you go," he said, handing me the jar. "You made that."

The honey glowed in my hands, like a living, breathing thing. It was warm, and I loved it because it made sense when nothing else did. It was a pure example of what Grandpa had been trying to explain inside the bus—that beautiful things don't come to those who simply wish for them. You have to work hard and take risks to be rewarded.

But he wasn't exactly correct when he said I alone made the honey. Both of us harvested it, but the bees made it. They'd collected nectar from millions of flowers to make this one small pound of honey in my hands.

All of us, humans and insects, in our separate ways, had traveled far, navigated dangers and labored to exhaustion for a shared obsession.

We made this honey because we believed we could.

9

Unaccompanied Minor

1977

The summer after I turned seven, a letter appeared in the mailbox addressed to me. Granny read it first before handing it over.

"Your father wants you to visit him and his new wife," she said. "You don't have to go if you don't want to."

I hadn't heard from Dad since we said goodbye in the driveway two years ago. I unfolded the crisp pages and held them to my chest, tracing the imprints where Dad had pressed the pen to paper, as if to convince myself that his hand had really made the marks, and that he had written these words specifically for me. It was physical proof that Dad did love me after all. Granny and Mom almost had me believing that Dad was gone for good, but now I had evidence showing they were 100 percent wrong. I believed my luck had finally changed, and now good things were finally going to start happening to me. Not only was I finally going to see Dad again; now I had a second mom. Grandpa had explained that *step* meant you got two of something. Could it be, that like the bees, I was getting a new queen to replace the one who was failing?

"I want to go," I said. "Matthew, too?"

"He's too young to fly alone. Airline rules."

Granny frowned as she stuffed the letter back in the envelope, and I couldn't tell if I had permission to go or not. She sat there for a moment, tapping the corner of the letter into her palm, thinking all this over.

"Let's talk to your mother," she said.

Mom sat up in bed, scanned the letter with a blank expression, and then let it drop from her fingers and flutter to the floor. She picked up her paperback murder mystery and resumed reading, as if Granny and I weren't in the room. A few seconds passed, and she lowered her book and peered at us over the top.

"You two can go now," she said in monotone.

"Sally..." Granny said in the soothing voice she reserved for calming elementary school students. She took a few steps toward the bed.

"I SAID GET OUT!"

Granny jumped back and put her hand over her heart, then shooed me out of the room, softly closing the door behind her with a click. I could hear Mom's muffled sobbing, and knew that my trip planning was indefinitely postponed. I retreated to the living room, cranked the TV and disappeared into the canned laughter of a sitcom, slamming my senses with forced cheerfulness. I was determined to see my father, no matter how much Mom cried about it, and I refused to let my visit get lost in her sadness. Mom's mood could suck all the energy out of the room, leaving everyone around her weary and hopeless. Now that Dad was reaching for me, I couldn't let Mom ruin it.

Eventually it was decided that I could go. There was no direct discussion with me about it; just one day Granny let me know she had written to my father to make arrangements for me to visit for one week. In the days leading up to my trip, Mom became increasingly anxious. She tossed and sighed in her sleep, her mind racing with a growing list of things she wanted me to retrieve from Dad's house.

"Hey, hey, you awake?" she'd whisper in the middle of the night.

I would try to fake-snore, but then she'd shake my shoulder, just a little.

"Meredith."

"Hmmmm?"

"Make sure you get my Bobby Darin records. And the Kingston Trio. Those are mine, not his."

I was drowsy, but I knew she'd remind me many more times, so I didn't answer. She poked me again. "Did you hear what I said? Repeat it back to me."

"Bobby and King Tree," I mumbled.

In a flash, she reached under the blankets and flipped me around to face her. A cymbal crash of adrenaline woke me, and when my vision registered, her face was just inches from mine. She seized my shoulders and spoke slowly, pronouncing each syllable.

"Bob-by Darr-in. King-ston Tri-o."

Her grip was strong, too strong, and the desperation I

could feel in it gave me the willies. I repeated the names just so she'd release me. She let go, and I wriggled to the opposite edge of the bed out of her reach. But her voice still found me in the dark.

"Don't forget the gold baby bracelets. Now listen—there were two. One is yours and one is Matthew's. Your names are inscribed in them. I know he has them. If he says he doesn't, he's lying."

I said I would, only to appease her. I didn't care about any of this stuff, and I didn't want to ask Dad for it, and I resented her for taking my trip and turning it into hers. But I knew there would be hell to pay if I didn't follow her instructions. Each night, her list grew. She wanted the pearl necklace and matching teardrop earrings she wore on her wedding day. The framed Sears baby portraits of Matthew and me. A wool coat that had belonged to her grandmother. She hovered as Granny helped me pack, pulling some of my clothes back out of the white suitcase to ensure there was enough room for her things. Worried I wouldn't remember everything, she wrote a list of her possessions and pinned it to the orange lining of the suitcase.

When my plane ticket arrived in the mail, Granny ripped open the envelope and examined it closely for the price. "If he can afford this, the cheapskate can pay more child support."

She settled before her writing desk, opened a drawer, and whisked out a thick piece of cream-colored stationery. I heard her sentences scratch out in a prosecutorial fury. Occasionally she held the letter up and examined her prose, reflected for a moment, then slapped the paper down to strengthen her arguments. Once she was satisfied, she licked the envelope and added the letter to my suitcase.

I didn't let myself get visibly upset by all the errands Mom and Granny were giving me. And once I was above the clouds on my fourth free 7-Up, it was incredibly easy to forget all about their notes in my suitcase. On my right shoulder I wore a required sticker that read, "Unaccompanied Minor," which I quickly figured out meant that I was lavished with attention from stewardesses bearing snacks and toys. The pretty ladies checked on me constantly, wanting to know if I wanted pillows or more crayons or if I'd like a pair of silver wings pinned to my denim jacket. I was the only kid alone on the plane, and that made me interesting to the other passengers, who asked me a lot of questions about where I was traveling. I was so excited to see Dad that I eagerly explained, but I didn't always get the response I expected. Some of the adults were delighted when I told them I was on my way to see my father; but others gave me a pained smile and changed the subject.

When the plane touched down, a stewardess instructed me to wait until everyone exited before I could get out of my seat. Those were the rules for children flying alone, but it was terrible torture. Time seemed to go backward as people fussed with their coats and bags while I bounced in my chair, silently pushing them down the aisle with an imaginary snowplow. Finally, my keeper materialized, took my hand and led me off the airplane. The airport was teeming with people, so many arms and legs blocking my view that I couldn't look for Dad. I clutched the stewardess's hand, afraid I'd get lost in the throng.

"What does your father look like?"

"He has black hair and he's tall," I managed, which didn't narrow things down much. It had been so long since I'd seen him, I wasn't exactly sure if I could pick him out in a crowd. She pointed to a stranger standing near a window with brown hair, and another chubby man sitting in a chair who was reading a newspaper. I shook my head no to both. She walked me toward the seated man anyway.

"Sir, is this your daughter?"

The man startled and lowered his newspaper. He shook his head and hid himself behind it again. I strained harder to see through the knot of people, but I couldn't locate Dad. We walked through the crowd once, twice, and doubled back for a third pass as my anticipation hardened into a stone in my throat. He forgot to come. Or worse, he remembered, but still didn't make it. He had changed his mind and decided he didn't want me after all. I braced myself for the moment when the stewardess would walk me back onto the plane to fly back to California. Granny was right. Dad was no-good.

I could feel the stewardess pick up her pace. The crowd was thinning, and she was running out of options. I wondered whether she could take me to her house. As my escort guided me toward a help desk, a man with a bowl haircut and a bushy mustache started walking toward us. The stewardess pointed.

"That him?"

The man was wearing what looked like a disco shirt with a wide collar. The fabric looked slippery, and had a print of black spirals bouncing over a maroon-and-green background. His tan corduroy pants flared at the bottom. My dad was the opposite—he had short hair and a shave, and always wore plain button-down work shirts tucked into straight slacks. This person was shaggy, more like a hitch-hiker. Or one of the Monkees.

"No," I said.

"Hey, kiddo."

The deep voice stopped me cold, and I instantly let go of the pretty lady's hand. The hitchhiker man tossed his bangs out of his eyes and grinned. "You must've walked right by me. I was standing here the whole time," he said.

I looked up and saw the sharp V of his widow's peak and knew it was Dad. I jumped into his arms and buried my face in his neck, inhaling a familiar scent of WD-40 and Old Spice. When I looked up again, the stewardess was gone. Dad kissed my forehead, tickling me with his mustache.

"You look different," I said.

"What, this?" he said, tugging on his mustache.

"Yeah, it's scratchy."

He set me down, and then stretched my arms out to either side to assess my wingspan. "I wasn't looking for a girl so tall."

I detected pride in his voice, and felt as if I had accomplished something incredibly grand, simply by growing. I was brilliant, I was miraculous and I was perfect beneath his approving gaze. As he led me through a labyrinth of bustling corridors, I sensed something inside me click back into place, a feeling of becoming whole again.

Dad drove a two-door Ford Mercury Monarch he called "The Beltway Banana." It was yellow inside and out, from the paint job to the upholstery, the steering wheel and even the seat belts. Its exuberant color amplified my already giddy mood. On the drive, Dad explained how to pronounce my stepmother's name: "Dee-ann." It sounded glamorous to me, a name belonging to a stewardess, most definitely. D'Ann had a big Italian family, Dad explained, with lots of brothers and sisters and cousins, and I was going to meet them all. Twenty or so relatives would feast at a long table in the middle of Nana Stella's kitchen, he said, eating spaghetti and cannoli until our bellies burst.

"And," Dad said, pausing for dramatic effect, "Stella always makes *three* desserts."

I had no idea Dad was having this much fun. I had been so busy missing him that I hadn't really considered what he was doing in Rhode Island. Now I could see that he had been rebuilding a family. But were these new people my family, too? I wasn't quite sure how it all worked.

"Did you get my letters?" Dad asked.

I told him I got the one with the plane ticket.

"What about all the others?"

"Others?"

Dad clenched his jaw and muttered what sounded like a curse word under his breath. I told him I didn't get any other letters from him.

"They must be throwing them out," he said.

Each day Granny drove down Via Contenta to the post office at the end of the block, opened a small door with the number 23 on it and pulled out the mail. She brought home bills and news magazines and letters from relatives and friends. If there were letters from Dad, I never saw them. Granny liked to say Dad was untrustworthy; but this made Granny downright dastardly. I looked down at the denim jumper and matching jacket I wore—a gift from Granny for the plane ride. I couldn't understand how the same person who took me shopping for a new outfit could also steal the most precious thing from me. My mind spun trying to find a logical explanation. Maybe the post office lost his letters. Maybe Dad made a mistake and sent them to the wrong address. Maybe Granny was just saving the letters for when I got older. Did Dad really write to me, or just say he did? Or maybe there were just too many secrets and lies flying back and forth for me to sort anything out.

"Why don't you call on the phone?" I asked.

"I've tried. Your grandmother hangs up on me."

I felt stuck. Granny, Mom and Dad were locked in a war that was bigger and stronger than me. My family was the opposite of a beehive. Instead of working for one another, all they did was conspire to make each other miserable.

Dad turned on the radio and a punchy jazz tune filled the car, the harmony gently blowing our bad mood away. He tapped his fingers on the steering wheel in time to the beat, and then informed me the saxophonist was Charles Lloyd, who lives in Big Sur. Grandpa and I rarely saw any other people when we went to his hives, and it seemed strange to think anyone else lived there. Especially a famous person.

"Does Frank still have his bees?"

I told Dad that Grandpa was teaching me how to be a beekeeper.

"I remember he took me inside that old bus once," Dad said.

"You've been in the honey bus?" I couldn't believe these two separated parts of my life were once ever together.

Dad got a faraway look and explained it was before I was born. "Your grandpa was always nice to me. Make sure you tell him hello from me."

I promised.

Dad now lived on the opposite side of Narragansett Bay in Wickford, a tiny colonial town with a main street of eighteenth-century brick buildings. We passed a harbor of sailboats gently rocking side to side, and turned into a neighborhood of simple one-story New England–style homes with painted shutters and screened-in porches. Dad parked before a faded blue house, and as we got out of the car, the screen door swung open and a petite woman bounced toward us, her dark hair swept back in a long ponytail. She was put-together in stylish clothes and matching heels, and she wore makeup and her nails were painted, and I immediately thought of my convertible-driving Fantasy Mom.

"I've heard so much about you," she said, wrapping me in a Chanel N° 5–scented embrace.

D'Ann held my hand and twirled me around to get a good look at me.

"You look just like your father," she said. She didn't pronounce her *r*'s, and the way she said "fah-thah" made me giggle nervously. But she laughed with me, like we were best friends sharing a private joke. "Who wants ice cream?" she said.

Just like that, she was approved.

When I stepped inside Dad's house, familiar objects triggered a dreamy sensation of walking back in time. I recognized bits of my old life, but in this new context I became uncertain of what I was remembering. There was the same black Naugahyde couch, but with an enormous black-and-white cat snoozing where Betty once sat and twirled my hair. A painted eagle on the headboard of a rocking chair looked familiar. Dad's reel-to-reel music player was in the living room, but now it shared space with an upright player piano.

D'Ann patted the space next to her on the piano bench, and I sat down. She folded back the music rack to expose the ivory keys, then slid open a door in the upper panel and inserted a scroll with perforated dots into the well. She put her feet on two pedals and pushed them one at a time, and the keys started moving all by themselves, banging out the Elvis tune "Hound Dog." My mouth fell open as a ghost tore it up on the keyboard, and I asked her to do it again, and again, transfixed. D'Ann swapped scrolls and then "Great Balls of Fire" filled the air. She opened a nearby closet to show me the top shelf was filled to the ceiling with more scrolls.

Thus began my week of living like a princess. I pretended that I was an only child with two happy parents who doted on me. I didn't even have to share the limelight with Matthew—a wicked thought, but I couldn't help myself. It gave me a rush to try on a different girl's life, and I inhabited my role so completely that Mom faded from my mind. Dad and D'Ann had planned so many adventures over the next seven days that there simply wasn't time to think about California. We had picnics on the beach, drove to the pick-your-own strawberry place and then stayed up all night making jam. D'Ann made a shirt for me on her sewing machine, and let me try on all her face creams. When the weekend came, D'Ann took us to her family home for a big Italian dinner. Her parents and siblings were boisterous, full of jokes and second helpings, filling my plate high, inviting me to foosball games in the basement, rides on the tandem bike and badminton matches. At the end of the night, my new aunts and uncles pressed folded five-dollar bills in my hand "for ice cream."

I grew so addicted to being the center of attention that soon I started forgetting my manners. Each time I asked Dad or D'Ann for something and got it, I was emboldened to push for more. There was a danger of becoming spoiled, but I couldn't resist experimenting with their devotion, testing its strength and durability. Each time the results came back positive, it was like a little hit of dopamine, a squiggle of joy at hearing the loveliness of that word *yes*. I encouraged them to dote on me because it pushed away my rising dread that this was all going to end, and soon I would return to a world that did not revolve around me.

One night, as the three of us watched a movie in bed together, Dad got up and asked if we wanted anything from the kitchen.

"English muffins! With butter!" I commanded, not taking my eyes off the TV.

D'Ann nudged me and pointed at Dad, who was in the doorway with his hands on his hips. "Shouldn't there be a *please* in that sentence somewhere?" he said.

I was mortified. I had forgotten who I really was. I had

become an insatiable baby bird; no matter how many worms Dad put in my mouth, I kept squawking for more. It wasn't even food I wanted; I was hungry to find out how far he'd go to indulge me. But I'd finally found his limit.

"Please," I croaked.

He nodded, and I sunk back into bed and pulled the covers over my head, hiding from his disapproval. I had almost lost Dad over some toast. I vowed to be more polite, to go back to the girl who kept her thoughts to herself.

I found Dad the next morning glugging down a tall glass of milk. He was wearing shorts and salt-cracked leather Top-Siders, and D'Ann was packing sandwiches into a cooler. It was one of those New England summer mornings when the air already felt milkshake thick; every piece of furniture I sat on stuck to my legs. Dad downed the milk and then put the glass in the sink. I wasn't sure if I was still on his bad side, so I waited for him to speak first.

"Let's go find a breeze," he said.

That's how I knew that everything had been forgiven.

The beach was the perfect place to spend my last day with Dad and D'Ann. It always seemed like time moved more slowly at the shore, away from clocks and telephones and schedules. I wanted to stretch our last hours, dreading the thought of having to say goodbye to my father yet again. I had an oversensitive reaction to letting go of him, because it reminded me of all the times we'd been torn apart against our will. I was afraid of that feeling I always got when we parted, a clawing feeling on the inside, like fingernails raking a line from behind my collarbone and through my insides, all the way down to my belly button. I was afraid of getting back on the plane without him. I didn't know if I was strong enough to handle it.

I pushed these thoughts aside as the blue of the ocean came into view. Someone must have called ahead and reserved the beach just for us—the parking area was empty save for the gulls circling overhead and the few lone surfers peeling themselves out of their wet suits. We walked along a boardwalk, passing a snack shop where a machine spun a cotton-candy web, and above it on a second story a riderless carousel whirred to old-timey piano music. We crested a dune, and spread before us was a glittering blue crescent, with steady curlers of foam rolling toward shore.

Dad reached the water first and splashed out to his knees, and I followed behind in his wake, squealing as if ice needles were piercing my skin. The water foamed around our legs, hissing as the undertow sucked the sand from underneath my feet. Dad clasped his hands above his head and arrowed himself through the belly of an oncoming wave, ducking under it and popping up on the other side to float on his back, his arms out in a T-shape for balance and his long feet cutting the water like shark fins. He made it look effortless, as if his body was made of Styrofoam. He lifted his head to address me.

"Now you!" he hollered.

I mimicked his entry and launched myself directly under the next rolling barrel of water. I blinked in the stinging salt water, and although it was murky, I could see phosphorescent bits of something floating all around me, like gold dust in the water. I kicked toward the light, and when I broke the surface, I felt arms encircle me from behind, and all of a sudden I was sitting in a throne made from Dad's bent knee and chest as he braced me from the next wave with his back.

He showed me how to float by filling my lungs with air and then holding my breath, and we bobbed like sea otters for so long that my fingers puckered into prunes, and eventually, growls of hunger emanated from my stomach. We rode the next wave in on our bellies, and joined D'Ann on the blanket for lunch.

"I was about to call the Coast Guard, you were out there so long," she teased. She handed us ham sandwiches, and tore open a bag of potato chips and set it in the middle of the blanket. Dad chomped his sandwich, consuming it in four bites. Then he stretched out on his back, propped his head up with a towel and placed a hill of potato chips on his stomach. He crunched loudly and let out a long, satisfied sigh.

"Can't believe I have to go back to work," he announced to the bluebird sky, which I think was his way of saying he didn't want the week to end, either.

I dug at the sand with my toes.

"Me, too," I said.

D'Ann reached out and silently rubbed small circles on my back. We finished our lunch in silence, chewing slowly, and I tried to not think about tomorrow.

That night, Dad tucked me in as he had all week, but he sat with me longer than usual. He flicked off the light, and the bug zapper outside the window cast a purple glow into the room.

"I wish you didn't have to go," he said, pulling the sheet

up to my chin. He sat back down, and the springs squeaked under his weight. I could hear him scratching his scalp, a nervous tic.

"So, do you like living in California?" he asked. In the darkness, his words sounded heavy and important.

A big moth flew into the zapper and sizzled.

"I mean," he continued, "are you happy?"

These were big questions that I'd never been asked, and I wasn't sure what kind of answer he wanted. I had never considered my own happiness, so the question took me by surprise. I wasn't happy like those kids who goofed around in music class, but I wasn't sad like Mom, either. I was somewhere in the middle, but was that where I was supposed to be? I wasn't sure, so instead of answering I pulled at a loose thread on the sheet.

This was the serious conversation we'd been avoiding all week. Both of us had been reluctant to interrupt our vacation with reality. Now his words were ruining the spell, reminding me that this week as his full-time daughter had been nothing more than make-believe.

Dad tried again.

"Is your mother nice to you?"

Nice wasn't the right word. Mom was Mom. She wasn't nice; she wasn't mean. She wasn't anything, really. I tried to come up with the right description, but I couldn't figure out how to put her into words. He must have thought my silence meant I was hiding something. He lowered his voice to barely a whisper.

"Does your mother ever...hit you?"

I bolted upright in bed, suddenly not liking where this

conversation was going. The question was preposterous. She would never do that. "What? No!"

An uncomfortable silence stretched out between Dad and me. I still hadn't told him about Mom's list, or Granny's letter, not because I was hiding them, but because I'd truly forgotten in the hurly-burly of our week of fun. He scratched his scalp again and said he was glad that everything seemed okay in California.

"But you know you can always tell me anything, right?"

It seemed like a good opening to bring out Mom's list. I pulled the suitcase out from under the cot, found Granny's letter inside and handed it to him.

"From Granny. She wants more money."

He crumpled the envelope without opening it, and chucked it at the wastebasket next to his desk, missing it.

"Her letters are so nasty, I can't even read them anymore."

I showed him Mom's list of items to be returned. Dad set the paper down on the bed and cleared his throat.

"Would you rather live here with me?"

His offer shimmered in the dark like the tail of a comet. Pretty, yet out of reach. After a week of entertainment, my insides screamed yes, but talking secretly with Dad about leaving Carmel Valley seemed like I was being sneaky, somehow. I couldn't leave Matthew behind. Grandpa wouldn't have help with his bees. Daughters weren't allowed to walk away from their mothers, were they? The idea felt sinful. Dad's offer was tempting, but I felt I didn't have the right, or the power, to switch parents. When bees make decisions about places to live, the whole group decides together. They spend days inspecting possible homes, and they vote by dancing, deciding together when to swarm and where to relocate. They discussed it first, listening to everyone's input. I would get in trouble if I made this decision alone, wouldn't I?

"You don't have to go back," he said.

Dad anxiously twirled his wristwatch as he waited for me to answer. I felt compressed by the enormity of the decision, as if I couldn't get enough oxygen into my lungs. I knew that I could never tell anyone that Dad had made this offer. I worried it was a deceitful idea, yet if he asked again I would say yes. I feared what Granny and Mom would do if I didn't come home. I struggled with the impossibility of what I wanted but couldn't have—for all the adults in my life to get along. I was suffocating in indecision, and wanted Dad to make the decision for me.

When our silence became unbearable, I whispered to Dad that I was okay.

I said that I wanted to stay in California. I lied and said that everything back home was fine; that Mom was fine. It was what I knew best, so I chose it, even if it meant returning to Mom's broken heart. But as soon as I chose, I regretted having made a decision at all.

"Well, if you ever change your mind, you can live here with me, you know," he said, kissing my forehead.

He shut the door, and I stared at the purple patterns on the wall from the bug zapper, wondering if I had made a terrible mistake. When I finally fell asleep, I had a nightmare that a cackling witch was squeezing my waist with her long, bony fingers, breaking me in two.

It was still dark when Dad shook me awake. D'Ann

waved from the driveway as we pulled away from my seven days of some other girl's life. Dad stopped at a doughnut shop, and I inhaled three glazed doughnuts in a row, not even really noticing how they tasted.

"I'll see you next summer. Matthew, too," he said.

"That's a long ways away," I said.

We couldn't think of anything else to say for the rest of the ride, already feeling our separation before it came. When it was time for me to get on the plane, Dad had to pry my arms from his neck. Another doll-like stewardess appeared out of nowhere and took my hand. I knew the drill by now, and let her place the sticker on my shirt and lead me away as I used all my willpower to not look back.

She didn't let go of my hand until I was buckled in my seat, and as soon as she was gone, I buried my head in my hands and keened. I was the kind of sad that didn't care if other people were looking. I ached for my father even more than I thought possible, because now I knew what I was missing. But to be with him, I'd have to give up my life back home, and I didn't want to do that, either. I wanted to stay with him, *and* I wanted to stay in California. I wanted both, but both wasn't a choice. I didn't know if I had made the right decision, and I wanted someone, anyone, to tell me what I should do. Trying to figure this out felt like being pulled apart down the middle, with Dad tugging on one arm and Mom on the other. I tried concentrating on the happy parts of my trip—the player piano and the bowls of spaghetti with my new Italian relatives—but knowing I could only borrow them made me cry even harder.

The stewardess returned and knelt down in the aisle

and handed me a tissue. She patted my arm and told me everything was going to be all right. I looked away from her stupid promises. She didn't know me, and she didn't know what was wrong, and she was just saying that because I was making the other passengers uncomfortable. I kept crying at full tilt, ignoring the coloring book, crayons and Cracker Jacks she placed in my lap. I wept until my nose was so stuffed up I couldn't cry anymore. I leaned my head against the round window, closed my eyes and wished the heavens above would swallow me whole.

I slept fitfully on the flight, cycling through a pattern of waking up, wondering where I was, remembering, and going numb all over again. By the time the plane touched down, I was cranky, hungry and a bit more skeptical of the us-versus-him family tree Mom and Granny had drawn for me.

Granny was waiting for me at the airport gate, and to my surprise Mom was standing by her side. I took this as a sign that Mom must have missed me. I relaxed a bit; perhaps California was the right decision. We made small talk on our way to the car, me answering that the weather was nice and that the trip was good. Yes, it was "nice" to see my father.

"That's nice," Granny said.

We had a two-hour drive ahead of us, and I stretched out on the back seat, while Granny started the engine. Mom snapped on her seat belt in the passenger seat. Then she turned around to face me.

"What does she look like?"

It took me a second to figure out whom she meant.

"I don't know. She has dark hair."

"What do you mean, 'you don't know'? Is she prettier than me?"

I picked at my fingernails rather than tell the truth.

"How old is she?"

I told Mom that I hadn't asked.

"Well, would you say she looks younger or older than me?"

I turned my head away and stared at the ceiling.

"Meredith! Did you hear me?"

I tried to say I was tired. I tried to fall asleep. Granny drove in silence as Mom interrogated. I tuned her out until her words were a blur, and I transported my body back to Stella's house, with the marinara sauce bubbling on the stove and Grandpa Duke cracking open a beer and talking about his golf game as Dad pretended to be interested in the sport. Uncle Roland was in the driveway, patching a hole in his canoe. Uncle Jeff was pushing me in the tire swing. A football game was on in the background.

Mom wanted to know if I had retrieved everything on her list.

Granny kept her eyes on the road. "Answer your mother," she ordered.

I mumbled that I only had the baby pictures. Mom wrinkled her face as if she had just sniffed a carton of expired milk.

"Meredith, dammit! I told you! One simple thing. You can't even do one simple thing!"

Mom and Granny argued back and forth over whether I should call Dad when we got home and demand he mail the missing items, or if Granny should write him another letter. Mom wanted me to call immediately and let her listen in. Granny talked her back down, and they finally agreed to try a letter first. Their side conversation gave me fifteen minutes in peace. Then Mom turned her attention back to me.

She asked me about Dad's new house. How big it was. What kind of cars they drove. Did D'Ann cook? What did she cook? I gave one-word answers, which only angered her more. She threw up her hands.

"What did you do, sleep the whole time?"

I told her we went to church on Sunday. Mom snorted in disgust.

"She's Catholic, isn't she? What does her family think about all this? Divorced Catholics aren't allowed to remarry, you know!"

I lost my patience and kicked Mom's seat from behind. "I don't know!"

Granny finally intervened. "Meredith, you do NOT talk to your mother that way!"

I kicked Granny's chair, too. "Dad says you threw out his letters!"

Now all of us were screeching like monkeys in a cage.

"I have done no such thing!" Granny said. "How dare he say that!"

Somebody wasn't telling the truth, but I didn't care anymore. I was exhausted from trying to figure out what to do with my life, and I just wanted to go to sleep. Mom continued questioning me all the way to Carmel Valley, asking the same questions different ways, trying to trick me into answering. But when we reached the house, she got to the question she really wanted to ask. Her voice became suddenly soft and small, like a young child's.

"Did your father ask about me?"

I hesitated, and finally said, "No."

She sank into the passenger seat, defeated.

I stepped out of the station wagon and noticed that the door to Grandpa's office, on the left side of the carport, was open. I found him at his desk, the papers pushed to one side, hunched over his handmade redwood jig, assembling new wooden frames for the hive and spooling wire through them. The horizontal wires acted as a brace to hold paper-thin sheets of wax that were stamped with hexagons to give the bees a foundation from which to build new honeycomb. He heated the wires using a contraption he'd fashioned from a disassembled light bulb socket, then pressed the thin sheets of wax onto the hot wires to fuse them in place.

"You're back!" he said. "Good. Gimme a hand here," he said, handing me a pair of wire cutters. "Snip that wire right there for me, would ya?"

Grandpa's office smelled of hot beeswax, dust and after-shave. I inhaled and suddenly felt calm again. I didn't realize until that moment how much I had missed him. I missed the bees and our trips to Big Sur.

He handed me a finished frame, the signal to give him an empty one, which he placed on the jig and continued stringing wire. He asked me if I had had a good time, and I told him about the beach, my stepmom, about all the ice cream I ate. I mentioned that Dad said hello. I told him everything that I didn't say in the car ride home.

It was a relief to talk about my trip with someone who was really listening. I asked him how the bees were, and he said he'd been busy catching swarms. Three of them.

"One was really high up in the rafters of a house," he said. "I could have used you to hold the ladder."

"Where did you put the swarm?"

"In the backyard, with the other hives."

He looked up from his work and saw what I was going to say before I said it. He set down his tools, stood and hiked up his sagging pants, and reached for my hand.

"Okay, let's go look at 'em."

His hand engulfed mine, and I felt his calluses pressing into my palm, and knew that this was the right choice.

10

Foulbrood

1978

Mom called out to me from the bedroom. She did this sometimes when she wanted more water or aspirin, so that's what I figured it was—she needed me to bring her something. But when I came in, she was rummaging around in the closet, pushing boxes and sweaters aside on the top shelf. She pulled out a board game and handed it down to me.

"I need you to play with me," she said, getting back into bed and reaching for the box. She took off the lid and pulled the game out.

"What is it?"

"Ouija," she said, pausing to take a long suck on a Fresca soda, while simultaneously balancing a cigarette between her pale fingers.

She patted the bedspread, indicating for me to sit next to her. She placed the board in front of us, and I saw it had a moon in one uppermost corner and a sun in the other. In the center was the alphabet in a western font, curved above a row of numbers. On the bottom were the words *No, Yes, Goodbye.* Oddly, there were no cards, no dice and no playing pieces. This game looked like it would be exceptionally boring.

"How do you play?"

"You use it to communicate with spirits," Mom said. "Like my dead granny."

It took me a second to absorb what she'd just said. She wanted to talk to the ghost of her dead grandmother. With me. I had no interest in poking around in the afterlife, because everyone knows ghosts don't like to be bothered, and they have the upper hand when it comes to revenge. But Mom wasn't joking. Her instructions were matter-of-fact, as if she really *believed* she could do this. At some point while studying the star charts in bed, she had advanced from astrology to séance. I did not see this coming. I wondered if maybe she had been in her room so long that she was starting to make up imaginary friends. I didn't know what to say.

"I had a granny, too, you know," she continued. "I loved my granny—she was the only one who was ever nice to me. Too bad you never got to meet her." A wistful expression flickered across her face. "She died just before you were born."

She tapped the ash off the tip of her cigarette into the ashtray and held up a spade-shaped piece of white plastic with a small round window embedded in it.

"You and I both have to put two fingers on this. Then you close your eyes and stay very still. It will start to move when the spirits want to say something. They will spell it out."

This did not sound very sane. But Mom was inviting me to do something with her and this was rare, and possibly a sign of improvement. Despite my trepidation, I put two fingers on the planchette next to hers. Our fingers touched, and it felt like a little embrace, a more intentional gesture of love than when she spooned me groggily in bed. We waited like that for several minutes, both hands on the plastic reader, staring at it, willing it to move. It felt nice to sit this close with my mother, and I really didn't care if the reader moved or not. She had asked me to be with her, and that was enough.

Finally, I felt the tiniest vibration beneath my fingertips.

"Are you moving it?" I asked.

"Shhhh, I'm making contact. Granny, are you here?"

The planchette picked up speed and swung in an arc, stopping over the word *Yes*.

A shiver coursed through me. I was certain that I wasn't moving it, and if Mom wasn't either, that meant some invisible presence was really taking control of the disc. I went limp just to be extra sure I wasn't moving it by accident. I could hear Mom's breath pick up.

"Do you have a message for me?" she whispered.

The reader sliced back and forth across the board, so fast we had to lurch to keep up with it. Mom bent over the board to make out the letters through the clear circle on the reader, sounding them out one by one to decode the message.

I M-I-S-S Y-O-U.

My stomach flipped, and I suddenly felt like I had to pee. Somehow Mom's dead granny was really talking to us. In less than five minutes, our innocent game had veered into the occult, and I suddenly felt like I was trapped inside a scary movie. I held my breath and checked the room for supernatural signs. I was so spooked that everything made me jump. What was that movement behind the curtain? Did I hear a footfall near the door? Was that a cold breeze, or was it dead Granny floating through the bedroom? I wanted to flee but was too petrified to move. The reader paused on the board, as the presence waited for the next question. Mom sat up and squeezed her eyes tight in concentration.

"Will I find another husband?"

The white disc didn't budge. She asked the same question, six or seven times more. Nothing. Whatever entity was in the room just moments before had clearly crossed back to the other side. So much for that. Ouija was a dud, I decided.

But Mom wasn't ready to give up. She remained hunched over the board, with the resolve of a person who was not going to quit until she got some answers.

That's when I really became truly scared. Worse than a ghost was the realization that my mother might be losing her marbles. She believed Ouija was real. She needed this cheap dime-store oracle to assure her things were going to turn around.

It made me feel sorry for her, to watch her beg the air for a man to make her happy again. She was pleading to the universe, to skeletons, to nothingness, for some small measure of hope. She seemed to be getting more desperate since my summer visit to Dad's, as if it had only intensified her feeling that she was stuck in place while life moved on without her.

Mom and I kept waiting, but the plastic reader didn't respond. Mom asked Ouija again, louder this time so the phantoms could hear her. When no answer came, she resorted to bargaining.

"Okay, how about just a boyfriend? Will I get a boyfriend soon?"

We waited some more. My arm was asleep now, and it felt like an anthill burst open at the top of my shoulder, sending an army of insect legs scurrying down to my fingers. Finally my fingers slipped off and knocked the reader to the right.

"Wait! It was moving just now, moving toward *Yes*." Mom lunged for my hand and placed it back in play. When the reader remained motionless, she compromised.

"I'm going to take that as a yes. It was moving toward *Yes*, you saw, right?"

"Definitely," I said, rubbing out a cramp in my forearm. I heard Grandpa start his truck to let it warm up, and I stood up to go. We had plans to go check the bees down the coast.

"Not yet!" Mom shouted, yanking me back to the bed by my wrist. Her grip was too tight, too urgent, and pinched my skin. It had a trace of roughness to it that was unsettling.

"Ow, Mom, you're hurting me."

"Sorry," she said absentmindedly, without looking up from the board. "Just a little more. Five more minutes."

I rubbed at the redness on my wrist where her fingers had briefly handcuffed me. I had no choice and I knew it; I had to keep playing until she dismissed me. I was trapped inside my mother's crumbling mind. I heard Grandpa rev the truck engine, and worried he might have to leave without me.

"About this boyfriend...will he be rich?"

This time I cheated and pushed the reader. Fast and hard over the word *Yes*. I think both of us knew what I had done, yet neither of us said anything. But I had to get out of the game somehow, because Mom was going to force the ghosts to tell her what she wanted to hear, no matter how long it took. So I came up with a white lie that we both could accept.

Mom's face relaxed as she put the game back in the box. She handed it to me and I put it back in the closet, burying it under sweaters where I hoped she'd forget it. By the time I turned around, she was napping with a smile on her face. She was content, knowing that good days were just around the corner.

I found Grandpa sitting on his tailgate, picking the mud out of his boots with a hive tool.

"I almost thought you forgot," he said.

"Mom wanted help with some fortune-telling thing."

Grandpa tilted his head to one side. "Come again?"

"Wee-gee."

"Never heard of it."

"It's not as good as cribbage," I said, referring to Grandpa's favorite game. He was teaching me how to play, using matchsticks for pegs and a piece of wood that he'd drilled with holes for the board. He smiled at my assessment, then opened the passenger-side door of his truck and waved me in with a dramatic bow just like a chauffeur.

When we reached Big Sur, the sky was a shock of pink

and orange above the low morning mist that had not yet pulled away from the coastline. The earth was damp beneath our feet as we made our way toward one of his smaller apiaries at the Grimes Ranch. Grandpa cut a path through a wildflower meadow, and I followed behind with the smoker and our bee veils. This group of hives was the easiest of his bee yards to access, clustered in an empty pasture with a view of Highway 1 and the Pacific. Long ago, one of his cousins who lived at the ranch started beekeeping, but his interest lasted less than a year. The cousin asked Grandpa for advice a few times, which turned into lessons, which turned into bee-sitting, which eventually became a full takeover of the hive. In the intervening years, as bees are wont to do, the colony multiplied, and on this day Grandpa and I were walking into a little clearing with twenty-eight beehives just beginning to hum with the sun's first rays.

The summer nectar flow was dwindling now, and the nights were coming sooner and colder. The late-autumn harvest would be smaller than the gangbuster summer crop, and Grandpa had to be more careful about how much honey he took from his hives so the bees had enough to eat until the flowers returned in spring. Once it got really cold, his colonies would sit out the winter, huddling together inside the hive and shivering their wing muscles to generate heat. The queen would be given the warmest spot in the nucleus, where she would slow her egg production and conserve energy. When the bees on the outermost edge of the cluster got too cold, they'd crawl inward to thaw, pushing other bees to the exterior, all the bees rotating and taking turns to keep everyone warm. It wasn't hibernation, exactly, it was more like a slowdown, the bees venturing outside only to relieve themselves or fetch water. The colony planned ahead for this, Grandpa said, by storing large amounts of pollen and honey in the frames closest to the hive walls, where their winter pantries could serve double duty as nutrition and insulation. Grandpa knew the personality and foraging habits of each colony, and which hives could afford to spare honey, which should be left alone and which would starve if Grandpa didn't feed them.

The hungriest hives got a sticky pollen patty Grandpa bought from the Dadant beekeeping supply catalog, made from pollen and brewer's yeast that came in flat pancakes the color of peanut butter pressed between waxed paper. He set the patties over the tops of the brood frames where the nurse bees could devour them quickly without having to travel far. Other times Grandpa mixed equal parts water and white sugar, and fed his bees sugar syrup by pouring it into in an old mayonnaise jar, hammering holes in the lid with an awl and then inverting the jar into a wooden block he cut to slide into the hive entrance and serve as a feeder. There was a space cut into the block to allow the bees in to lick the drips that fell from the jar. His third option was to take frames of honey from abundant hives and swap them into hives with paltry honey stores.

Our mission today was to open all his hives and redistribute frames of honey from the strong hives to the weak, and if any honey was left over, we'd take it back to the honey bus for ourselves.

As we approached his bee yard, a flock of birds vaulted from the ground to broadcast our invasion in their own languages: Chickadee, Bushtit, Warbler, Blue jay. All those wings at once sounded like the flags at my school on a windy day, and I stopped for a second, just feeling the sonic power of their collective outburst. Grandpa and I watched them soar toward Garrapata Canyon. When they were out of sight, I looked to the ground to see what the birds had found so interesting.

I felt something crunch under my shoe, and discovered I was standing in the middle of a bee battlefield, the ground littered with expired drones. Some of the male bees weren't quite yet dead, and dragged themselves in aimless circles through the carnage, toppling over every few steps on legs that were broken or lame. One pitiful drone was trying to get back into his hive, but kept getting pushed back by the bees guarding the entrance. Two bees attacked him, each one biting and pulling on a wing until the trio tumbled to the ground and continued wrestling. I watched aghast as they bit off one of his wings, and one of the guard bees airlifted the feeble drone, carrying it up and away in its clutches to unceremoniously drop it several yards away from the hive.

Grandpa must have seen the drones, but he stepped indiscriminately, smashing them underfoot as he went about the business of getting ready, lighting the smoker and putting on his bee veil, as if nothing was amiss. I tugged on his sleeve and pointed at the catastrophe on the ground. He glanced down, then handed me the smoker. I was careful to grab it by the bellows, where it wasn't hot.

"Winter's coming," he said. "Not enough food to go around. Time for the ladies to kick out the men."

Just then a wasp homed in like a jet fighter, landing its smooth, streamlined body on the back of a fuzzy drone that was struggling to stand. The wasp bit the drone's head off in two quick moves and devoured the eyes while its headless body continued to twitch. I grimaced and asked Grandpa why the bees had suddenly turned cruel.

Drones get pushed out of every hive, every year, he explained.

"Fewer mouths to feed," he said.

The drones try their best to fight back, but a hive has tens of thousands of female workers and only hundreds of male bees, so the fellows don't stand a chance.

"Remember how I told you the drones don't do any work? They just sit around and beg for food?"

I nodded.

"Well, now it's payback. If you're helpful, people will help you back. If you're only concerned about yourself, then... *skeeeeeech*!" He drew his index finger slowly across his neck.

"Jeez Louise," I said, parroting one of Granny's favorite expressions.

It's no big deal, Grandpa said, when it warms up again the queen will simply make more drones.

At that moment, I felt very, very relieved to be female. A hive was a matriarchy built on a basic principle of work and reward, but the sisterhood seemed to be taking their power a little too far. It didn't seem right, to kill your brother. Even if he was lazy. And I'd watched enough nature shows with Grandpa to know that all creatures needed both males and females to make babies. If a hive pushed out all the drones to die in the cold, how could the queen keep laying eggs?

Grandpa took my question and held it for a moment. He helped me secure my bee veil and lowered his voice: "Okay, smarty-pants, drones do have one job. To make the queen pregnant."

I set the smoker on top of a hive where it wouldn't catch the grass on fire, sensing a potentially intriguing story coming on. I listened carefully as Grandpa explained the cutthroat competition for the queen's affections. It all starts, he said, when drones pick up the scent of a virgin queen flying nearby.

"Like when a dog's in heat and the other dogs know?"

"Something like that."

He continued, using hand gestures, to explain that the drones soar into the air and gather into a cloud, getting ready for the virgin queen to arrow through them. When she leaves the hive for her wedding flight, she mates in the air with only the fastest and strongest suitors that can keep up with her. She couples with a dozen or more drones one after the other, and then returns to the hive with their sperm stored in her body. She spends the rest of her life laying eggs and fertilizing them herself.

Because a healthy hive can go for up to five years with the same queen, and hundreds of drones hatch and die each month, the math isn't in the drones' favor. Few ever get the chance to actually do the one thing they were born to do. More often, a drone is just an insurance policy, on standby in case a virgin queen suddenly flies by. But even if a drone does get his chance to mate, he won't survive the encounter, Grandpa said.

It was so quiet I could hear the waves hitting the rocky shore in the distance.

"How come?"

"His man part breaks off and he falls to the ground, dead."

"Gross!"

Grandpa looked taken aback. I could tell that my squeamishness disappointed him, that all this time spent in Big Sur country should have made me hardier, or at least capable of accepting the laws of nature. My outburst came from a soft, indoor kid.

"Gross? What's so gross about it? It's just part of life. If it's very quiet, you can actually hear it snap off. It makes a little popping sound."

I shuddered, ready for his story to be over. I grabbed the smoker and began sending puffs of smoke over the hive entrances to calm the bees. I blasted the guard bees with more smoke than usual, feeling the need to even the score for the drones. The bees scuttled back into the hive to get away from the odor of burning cow patty, which masked the banana scent of their alarm pheromone. Grandpa realized that he had lost my interest and pried the lid off one of the hives to peer inside at the honey supply.

We had purposely parked the truck several hundred yards away from the apiary and placed empty hive boxes on the tailgate. It's a little tricky to steal honey from bees, so we devised a system to outsmart them. First Grandpa removed a frame of wax comb solid on both sides with sealed honey, then he gave it one good shake to send the bees tumbling back into the hive. They took offense to this, and many returned to the air searching for their stolen property. The bees zoomed in mad circles around Grandpa's head as he flicked the returnees off the honeycomb with a crow feather, racing to outmaneuver them in a battle of wills.

When the frame was as bee-free as possible, he handed it to me and I sprinted to the truck, pursued by a handful of outraged guard bees. Once I reached the tailgate, I checked for stowaways on the honeycomb, and blew on them softly like Grandpa had shown me to irritate them just enough so they flew off. Once the frame was clear of bees, I slid it into the empty hive box and hid it under a sheet. The bees could smell the honey, and they would be back for it if we didn't keep it concealed. They would cling to it, all the way to Carmel Valley, and that would be their undoing. They could survive the trip, but our house was too many miles away for them to navigate back to their hive, and they would die alone.

The first two hives couldn't spare honey. Grandpa removed the top pantry boxes on the third, and then bent down over the box containing the nursery, his mustache practically pressed against the top bars, as if he were trying to dive inside. I came closer, and my nose picked up what he was smelling—a horrible stink like meat gone bad. Grandpa stood up and shook his head.

"Not good."

This hive was different from the others. When I placed my hand on the side of it, the wood was cold to the touch, without the usual warmth emanating from the colony's collective body heat. I looked down at the hive entrance and noticed very little traffic.

Grandpa took out a frame of honeycomb that was most definitely the wrong color. The wax was too dark, like coffee, and while it should have been covered with nurse bees tending to the brood nest, there were only a few sluggish nurses pacing over a rotting nursery, desperately looking for a healthy larva to feed. The wax seals over the birthing chambers were sunken and perforated, when they should have been smooth like a taut paper bag.

Grandpa plucked a foxtail out of the ground, and poked the stiff end into one of the wrinkled brood cells. When he pulled it out, a slimy brown string came with it. He examined the goo on the tip of the weed for a long time, as if he couldn't believe what he was seeing. He checked a few more cells, and they had all the same snot-like stuff inside where a white bee larva was supposed to be. Somehow the grubs had liquefied before they developed into bees.

"Foulbrood," he said. I heard defeat in his voice, and knew it was something bad. Something serious.

"Foul what?"

"A disease. Highly contagious. Only way to get rid of it is fire."

Grandpa stacked the hive back together, then took a pencil from his back pocket and drew a big X on the cover. I gasped, realizing that meant he'd have to burn it with the bees inside. Grandpa squeezed his forehead like he had a migraine, then ran his hand through his hair and looked off in the distance. He was sorting something out in his mind, so I waited a bit before I asked my question.

"How'd this happen?"

"The nurse bees fed food to the larvae that had a nasty bacteria in it. Destroyed their guts."

Grandpa could only guess where the bacteria came from. Could be anywhere, he said; a bee can pick up the bacteria from touching another bee, robbing honey from a sick hive, even from landing on a flower where a diseased bee had been. Developing bees get foulbrood when nurse bees feed them bee bread made from a mixture of nectar and pollen that has the bacteria.

"All I know is, it's nasty stuff. Can last for up to fifty years."

I watched Grandpa dismantle hive after hive and poke the brood cells with a dry weed. He moved methodically, more like a piece of machinery than a human being. By the time he was finished, a dozen hives had been doomed with an X. He would need to build a bonfire and burn them all together to keep the disease from wiping out the whole apiary. I watched him fetch a shovel from the back of his truck and, when he was a good distance away from the hives, begin digging a grave for his bees.

I had no idea that bees could get sick. In my mind bees were unstoppable balls of energy. Most died of exhaustion after six weeks, so they put every minute to use. Each day they visited thousands of flowers in a five-mile radius of their hive, stopping only when their tattered wings finally grounded them. Old bees were easy to spot; their bodies were thinner and balding, giving them a polished look. Now that I realized how vulnerable bees could be, I felt responsible for not protecting them. A good beekeeper was supposed to keep bees, not lose them.

Grandpa's pit was a foot deep, and he was standing in it when I finally approached.

"Are you going to do it today?"

"I'll have to come back tomorrow with gasoline," he said, as he stepped on the spade and plunged it into the earth. He yanked the handle toward himself to loosen the ground, then bent over and heaved a scoop of dirt off to the side.

I'd never heard Grandpa's voice sound so thin, and I wasn't sure how to be around him. I sat on the edge of the pit and waited until he'd spent himself digging. He took a seat beside me and dropped his head in his hands. I leaned into him and felt the warmth of his exertion. We stayed that way for quite a while, keeping each other company without talking.

"Well, that's that," he finally said.

"Are you going to lose a lot of money?"

Grandpa was looking out toward the horizon, and I wasn't sure if he had heard me.

"Money? You think I do this for money?"

His question made me feel like I was in trouble, but I couldn't figure out what it was I had done. I had disappointed him again with my wrong thinking, despite all his efforts to raise me right.

"Honey isn't what's important," he said.

I opened my mouth to protest but couldn't assemble a sentence. Why have a honey bus if he didn't care about honey? Everyone knows that honey is the absolute most important thing about bees. That's why they are called *honey*bees.

"Do you think the only thing a bee does is make honey?"

I knew a trick question when I heard one. So I carefully answered with a question.

"Yes?"

"Wrong. Bees make food grow," he said. "All the fruits and nuts on our trees. The vegetables in our garden."

Grandpa's grief must have been making him sentimental. I'd seen his artichoke bushes push up stalks taller than me and produce an artichoke with a punk rock head of purple hair on top—unassisted. The almond tree in our front yard made white flowers that eventually turned into green fuzzy pods, and then I watched those pods shed and leave behind woody husks with nuts inside. The tree did all the work.

"Plants make food," I tried to clarify.

"Not without bees, they don't," Grandpa corrected. "Flowers need to exchange pollen with other flowers to become food. Because flowers don't have legs, they need bees to carry their pollen for them. Pollen sticks on the bee when it flies from flower to flower, and there you have it. Pollination."

Without bees making pollen deliveries, Grandpa explained, many of the things in the produce section of the grocery store would vanish. I would lose my beloved cucumbers and blackberries. No more pumpkins at Halloween. Summers without watermelon. The cherries Granny likes in her Manhattans—gone. The world would be bland, and boring, and flowerless without bees, he warned.

Now it made sense why Grandpa was so distraught. Losing his hives was so much more than a personal disaster; it was a setback to nature itself. Not only would we lose produce, Grandpa said, but other animals would be in trouble, too. We needed bees to pollinate alfalfa and other grasses so cows and horses could eat. Mother Nature knit a careful plan in place, and if you pulled one thread of it loose, the whole thing could unravel. These insects that made most people run in fear were the invisible glue of the earth that held us all together.

Grandpa had just revealed a hidden staircase in my mind, showing me that there were so many things to learn, beyond what I could see with my own eyes. Before, when I looked inside a hive, all I saw were bees going about their chores, never imagining that their labors had anything to do with me. It was astounding to realize that every creature, no matter how small, helped keep everyone else alive in a hidden organization. If something as seemingly insignificant as a bee was silently taking care of us, what about an ant, or a worm or a minnow? What else didn't I know about the unseen contributions that nature was making all around me? It made me think that the universe had a plan for me, and although I couldn't always see it or feel it, I had to trust that it was there. It just might be that my life wasn't random, or unlucky, after all. I considered this possibility for a moment, and for the first time in as long as I could remember, I felt a trickle of worry slip away.

All this time I thought Grandpa and I were the ones taking care of the bees. When all along, the bees were the ones taking care of us.

"I'm sorry you lost your bees," I offered.

Grandpa stood, put his fingers in his mouth and a piercing whistle ricocheted up Palo Colorado Canyon. He sat back down, and within seconds, Rita came bolting out of nowhere, hopped into his lap and licked his chin.

"Sometimes things get taken away from you," he said. "But you can't let it get to you too much."

The good thing about bees, he said, is that they multiply quickly. If we were careful and attentive to the remaining hives, he could build his apiary back up to size within a year or two. Bees can take many hits, but they tend to always come back, he said.

I climbed into the truck and sat Rita on my lap to wait for Grandpa as he loaded honey supers in the back bed. Given the late season and the foulbrood fiasco, the yield was paltry, only a handful of boxes to take home. I heard the tailgate slam and when he sat next to me, I was struck by how tired he looked, with wilted cheeks and worry lines forming deep grooves across his forehead. He took one glance over his shoulder at the apiary and the awful chore that awaited him, and we pulled away.

The sun was directly over the ocean now, glinting like diamonds bobbing on the surface. This time there were no stories for the ride home. Grandpa was somber, lost in his own thoughts. Rita left my lap and curled up in his, as if she, too, could sense he needed cheering up. She nudged his belly a few times, and then rested her head on it and yawned.

"I'll help you," I said.

"What's that?"

"I'll help you get your bees back," I said.

Grandpa broke into a wide smile, and his face was suddenly familiar again. He reached over and patted my knee.

"Thank you," he said.

I reached over and turned on the radio, twisting the dial until the air came to life with a Johnny Cash song I'd heard Granny play on the record player.

Grandpa started to sing, and leaned over to ask me how high the water was, Momma. I knew the answer: two feet high and risin'.

Grandpa sang the question again, and again, louder each time and I responded in kind, yelling out, three feet! four feet! We shouted along with Johnny when he sang that his hives were gone and he lost his bees and his chickens were all up in the willow trees.

I heard the sadness in that song for the first time, but in a strange way it made us both feel better. We weren't the only ones at the mercy of nature.

11

Parents Without Partners

1980

Matthew and I were still in our pajamas, sprawled on our bellies shouting prices at the television. Our weekend ritual was to watch *The Price Is Right* or *Let's Make a Deal* to witness regular people like us win prizes that would bring them never-ending happiness. We memorized the games so that one day when we were old enough to drive, we'd gun it all the way to Hollywood, get on the show and Make It Big—so big we could buy a mansion with so many rooms we'd lose track of them all. Every room would have a waterbed, too.

After several years of dedicated study, I could recite within pennies the going price for almost anything, from a Corvette to a bottle of Clorox. On the screen a schoolteacher was trying to guess the combined cost of a trip to Hawaii and a Jeep, and despite my vigorous coaching from the sidelines, she was way overshooting it. I was so focused on the television that I didn't hear Mom walk into the living room.

"Who wants to go bowling?"

We peeled our eyes away from the Showcase Showdown. Mom impatiently shifted her white faux leather purse from one shoulder to the other. It was discombobulating to see her out of bed during the day.

"What? Why are you looking at me funny?"

We were in the middle of our sixth year of living with Granny and Grandpa, and by now Mom had become more like an older sister, tolerating us when she had to, but mainly avoiding everyone with a restless impatience. Our father had held true to his promise, and flew my brother and me out every summer for visits, but Granny had taken over as our full-time caretaker, and in that way Mom was insulated from the drudgery of adulthood. She was still without work, without friends, without motivation to get out of bed. My brother and I were so unaccustomed to taking direction from our mother that at first it didn't register that she was inviting us somewhere.

"Bowling?" I repeated, still stunned.

She let out an exasperated sigh. Her skin was so pale that blue veins showed through at her temples and wrists. She was wearing polyester yellow pants with an elastic band to accommodate her waistline, which had expanded considerably since we'd moved in.

"That's what I said. I don't have all day, here. You kids coming or not?"

I felt like we should be asking Granny for permission first, or perhaps Granny should come along as chaperone in case anything went wrong. I was dubious, but too curious to say no.

The closest bowling alley was an hour away in Salinas, and during the drive Mom explained that she had recently joined something called Parents Without Partners, and we were going to a bowling party for people like her.

"Ladies without husbands?" Matthew asked.

Mom rolled down her window a crack and let the wind suck the ash off her cigarette. "Men without wives, too," she corrected.

I wiggled my eyebrows at Matthew and leaned toward him. "Dating," I whispered. I pretended to make out with the palm of my hand, kissing it ferociously until he broke into a spasm of giggles.

"What's so funny back there?"

Mom's eyes snapped at us in the rearview mirror. All she saw were two cherubs sitting on the back seat of the Gremlin. I discreetly pinched my nose to stifle a laugh from slipping out. "I need you both to be on your best behavior. Don't do anything to embarrass me."

We promised to be good, although I didn't understand how we could embarrass her by throwing a ball at some pins. I looked out the window and saw rows of spinach and strawberries flicker past, blurring together like someone was shuffling a deck of green cards. Salinas was flat, and the fields lined up in military formation, as if God had first drawn the city on graph paper before creating it.

When we got out of the car, the air smelled of manure fertilizer, overpowering Mom's Charlie perfume. Her hoop earrings bounced as she hustled Matthew and me toward the entrance, but as we got closer she slowed her pace. She stood before the glass door as if she had had a change of heart. Mom fixed her lipstick in the reflection and tucked a few wisps of hair behind her ears. She adjusted the waistband of her pants. She had started dieting recently, eating mostly grapefruits and cottage cheese following the advice of a celebrity doctor named Scarsdale.

"Do I look fat?" she asked, turning sideways in the window.

Her tummy pooched out, but her legs and her arms were still regular size so she looked a little like she was pregnant. Matthew and I did not say any of this. We assured her that she was skinny.

"You really think so?" She looked over her shoulder to try to see her backside in the glass.

We nodded enthusiastically.

She bit her lip and looked back at the Gremlin, like she was trying to choose between curtain number one and curtain number two. One choice held diamonds; the other a donkey. She sucked in her stomach and held her breath. Then she let it out again and frowned.

"You're not just saying that? You really think I look okay?"

Other kids were running into the bowling alley, swinging the door open wide and letting out a heady smell of french fries and greasy pepperoni pizza. She grabbed our hands and squeezed. "Now listen, you two, don't ask me to buy things because you know I can't afford it," she said.

Matthew and I promised. She pushed open the door, and I heard the hollow clatter of crashing pins followed by jubilant cheers. My mouth watered at the smell of cotton candy, and a bank of pinball machines called to me with blinking lights and perky chimes. After the clerk handed us our leather bowling shoes, Mom walked Matthew and me to a lane where a group of sullen kids sat on a curved bench made of molded orange plastic. These were the sons and daughters of all the partnerless parents, forced to play together and clearly wishing they were someplace else.

"I'll be over there," Mom said, indicating four lanes over where the adults were mingling. Her purse bobbed on her hip as she speed-walked away from us. A strike boomed from the neighboring lane, and a group of men cheered and raised their beer mugs. The player who had just thrown the ball did a little air guitar move and stuck his tongue out like KISS.

Matthew and I looked back at our new companions and found a half dozen pairs of eyes boring holes into us. One spat sunflower seeds onto the floor near my feet, which most definitely was on purpose. The one with the earring said something in Spanish, and his buddies snickered.

"Hi," I said.

No one answered. I could sense all of us were the kind of kids who sometimes just wanted to hit something. More pins smashed nearby in a thunderclap, and I startled. I tried to cover it up by faking an itch and scratching between my shoulder blades, then I casually walked over to the bowling ball carousel and reached for a red one, but a girl snatched it away first.

"That's mine, *puta*," she said, tilting her chin up like I'd seen the boys do at school when they were about to pick a fight. I didn't know what that word meant, but I could tell it was bad, something to do with poo. Defeated, I sat on the bench next to Matthew. I put my hand on his back and his muscles were rigid.

"Gonna play?" I asked.

"Yeah, right," he said, covering his ears to the exploding pins. He hated it here. I got up to give it another try, being careful not to take the red ball. Once Matthew saw me playing, he would probably want to join me. But as I approached the lane, one of the boys blocked my path.

"What do you think you're doing? This is *our* game." He pointed up to the electronic TV monitor hanging from the ceiling. "You have to pay for your own."

I flopped back down next to Matthew, who was starting to cry silently. I tried to shush him quietly, but the mean boys smelled the salt water of his tears and pounced. They mock boo-hooed with sissy voices, and I stood in front of Matthew so he couldn't see them, shooting invisible death rays with my eyes. Unimpressed, they kept sniveling and chattering in Spanish, overjoyed that they had the power to scare a little kid. Matthew hugged his knees to his chest and curled into a ball. And that unleashed my inner tiger. I walked over to the boys.

"Now you've done it," I said. "I'm getting my mom."

The tyrants were suddenly silent as I swiveled on my heel and marched toward Mom, not sure exactly what I was going to tell her. She was sitting before a panel of illuminated buttons that controlled the scoreboard, cheering for somebody on her team. She glowed in a happy way that I'd never seen, and I forgot, for a second, why I had come over to speak to her. It was like I was watching someone I didn't know, someone with so much laughter inside her that she passed it around to all her friends. I called out to her, and as she swiveled around in her seat, all the joy evaporated from her face.

"What's wrong? I can tell something's wrong."

I explained we had a bully situation going on at lane two. So bad that Matthew was crying.

"What do you mean Matthew is crying?"

"Those kids are being mean to him," I said. "And they won't let us bowl."

She jackhammered her cigarette out in an ashtray built into the console.

"Well, what do you want me to do about it?"

"We need money to play by ourselves."

Her hand whipped out and she grabbed me by the wrist, pulling me close. Her words came out like a hiss. "What did I say about asking me for money?"

"I know, but..." Before I could finish, she was on her feet. She slammed her purse under her arm and practically stampeded toward the kids' lane. I watched the mean boys' eyes widen as she approached, but she went straight to Matthew, leaned over him and shouted at the back of his head.

"Why are you crying!"

I felt my face go hot as twin flames of fear and embarrassment licked my cheeks. This was not how this was supposed to go. She was supposed to protect Matthew from being bullied. Now his tormentors wore expressions of vindication, quietly ecstatic that the runt of our group was being scolded by his own mother. It would be open season on Matthew the second she walked away. Sensing this, he crumbled in defeat, sinking farther onto his knees.

Mom turned toward me and shook an accusing finger.

"You two will NOT ruin this for me! We drove all this way, now you are going to stay here until I'm ready to go. Do you hear me?"

Matthew stopped holding back and sobbed openly. Mom yanked him off the bench by his arm, and he hid his face in his hands, trying to make both her, and the whole bowling alley, disappear. The men in the next lane put down their beers and turned to look. The crash of pins silenced. The Spanish-speaking kids held their breaths. The bowling alley became library quiet.

I ran.

"Where the hell do you think *you're* going?" Mom hollered. Pinball players twisted away from their games to watch the commotion. I found my legs taking me to the bathroom, the one place I could hide from the ruins of our fake family outing. I locked myself in a stall and crouched on top of the toilet seat, in the futile hope that Mom wouldn't see my shoes and somehow miss me. Then I heard her pounding footfalls and I squeezed my eyes shut, held my breath and cowered.

Mom stomped into the bathroom like a bull, and pushed open each stall door with a loud boom, searching for me. I saw several pairs of feet scurry out of the bathroom, and I cringed: I had a mother who was so scary that strangers ran from her. I wanted to tell those other girls that she wasn't going to hurt them, that they didn't need to flee like that. But as Mom slammed her way down the row, I had another horrifying thought: maybe those girls had good instincts. I was the dumb animal that had cornered itself without an escape plan.

I saw the top of Mom's head when she stopped before my stall. A vein was pulsing in her forehead. She pounded on the metal door, sending tinny vibrations through the walls.

"Meredith, I know you're in there! You get out here right this minute!"

She flung her arm over the top of the door and grasped for the sliding lock with desperate, clawing fingers. It was safely out of reach.

"You need to OPEN this right now!"

She clasped the top of the door with both hands and shook, trying to wrench it open as I watched the flimsy latch strain against her force. My nerves sizzled as I tried not to think about what she would do if she reached me. She slapped the door again, and I flinched. Granny and Grandpa were far away and couldn't save me. I hugged my knees tighter and told myself it was just a bad dream.

"Answer me!" Mom roared.

I opened my mouth, but there were cotton balls lodged in my throat. It was dry like that time I got tonsillitis and all that came out was a weak rasp. I wanted to scream for help, but I was too ashamed to beg strangers to rescue me. It was just Mom; she wouldn't really hurt me, would she? I had never been afraid of her until now, and I wasn't sure what to do with the newness of that. She was scaring me, yes, but that was private information, not to be shared with polite society. I was frozen with indecision and whimpered helplessly.

Suddenly the stall stopped shaking. It was quiet for a few beats, and then Mom slammed her whole body at the door like a football player, trying to break it with her shoulder.

"Mom, stop," I whispered. "Please."

"What's wrong with you two kids?" she screamed. "Now BOTH of you are crying? You two need to grow up, that's what!"

She kicked the door.

"You don't call the shots around here. *I* do!" she said. Her breath was coming out fast, like she had just run a mile. Then I heard the flick of a cigarette lighter, and the crackle of burning tobacco as she inhaled. A cloud of smoke rose from the other side of the door. We stayed in our silent standoff for I don't know how long. Then I heard a man's voice.

"Ma'am. Excuse me, ma'am."

Mom regained her conversational voice. "You can't come in the ladies' room," she pointed out.

"Correct, that's why I'm going to need you to exit the bathroom unless I need to call the police."

"And who might you be?"

"Manager. Is somebody with you in there?"

I saw the cigarette stub drop to the floor and flatten beneath her bowling shoe. She sighed heavily and left me. I waited a few minutes until it seemed safe, then lifted the latch and crept out of the bathroom. Matthew waved me over from where he was sitting on a bench outside an office with windows. He pointed inside the room, where I could see Mom gesturing wildly with her hands, explaining something to the manager who stood with folded arms. I took a seat next to Matthew, waiting until the manager opened his office door for Mom and swung his arm wide, palm indicating the exit.

"C'mon, we're leaving," Mom said, taking one of our

hands in each of hers. We jogged to keep up with her as she hurried for the door.

"Happy now?" she said, as she slammed the car in gear, stomped on the gas and zoomed away.

We knew the question was rhetorical, and didn't answer.

"I could've met someone today, but you two certainly screwed that up good! That's the last time I take you kids anywhere."

I just wanted the whole day to disappear. I was sorry for being me; I was sorry she didn't have a husband and had to go to stupid bowling parties. I was sorry my brother was always the one getting picked on because he refused to fight. But most of all, I was just sorry that right now everything was wrong. Mom had emerged from the bedroom a different person from the one who went in. She had gone from a mouse to a mountain lion.

I closed my eyes and tried to remember what Mom used to be like before California. It was hard because I was a little kid then and now I was almost in middle school, and so much time had passed that I was forgetting Rhode Island things like snow and running through leaf piles and the lyrics to Beatles songs. Only a few memories of Mom were still clear in my mind—the bunny-shaped sheet cake we made for Easter once, decorated in white coconut with thin licorice for whiskers. I remembered watching the movie *Charade* in bed with her, trying to figure out where the treasure was buried. I can feel her hands on my back pushing me in a swing. There must be more.

When we got home, Mom was still fuming. She went back to the bedroom and got into bed, and we knew without being told that we should stay out of her sight. Matthew and I went outside to pick blackberries, and as we passed the honey bus on the way to the garden, we couldn't help but notice that the back door was ajar. We pulled it open and found Grandpa sitting inside with a five-gallon Wesson Oil tin wedged between his feet.

"Go find two small rocks," he said, as if he was expecting our company.

We returned, and he tied each rock to a piece of foot-long string. He dipped the rock and about half of the string into the tin of hot beeswax. He lifted it out quickly, held it until it hardened and submerged it again. With each dip, the candle got larger. He handed the wicks to us, and we copied his movements. We sat in silence together as the sunlight slanted into the bus, slowly making candles and pausing every once in a while for Grandpa to reheat the wax over the outdoor propane burner. Matthew's candle was starting to form a curve. Grandpa took it from him and rolled the taper in his palms to straighten it, then he handed it back. It occurred to me that I had never asked Grandpa how bees make wax.

"Little flakes come out from underneath their abdomen," he said.

"Whaaaat?" Matthew said.

Bees naturally produce wax flakes from their own bodies, Grandpa explained.

"Then they pull the flakes to their mouth, chew and mold them into honeycomb shape," he said.

Some bees are wax makers, and others are wax builders, Grandpa explained. When the bees get ready to build wax

Honeybees sculpting wax honeycomb.
(Photo by Kendra Luck)

When I was seven, Dad mailed a plane ticket and I flew by myself to Rhode Island to visit. We hadn't seen each other in two years.

Dressed as a hound dog for Halloween, complete with a pair of my granny's pantyhose on my head, with my ballerina-bodyguard friend Hallie at Tularcitos Elementary School.

A trip with our grandparents to
California ghost towns, 1983.
I am 13 and Matthew is 11.

Playing in the snow in the
Sierra Mountains, 1984. I am 14.

Just a small sample of the inventory Grandpa kept strewn around the yard
for his plumbing business. Stray cats loved rooting for mice in his piles.

My editor and I kept beehives on the roof of the *San Francisco Chronicle* building from 2011 to 2014. During this time I drove to Carmel Valley often to seek Grandpa's advice. **(Photo by Jenn Jackson)**

Queen cells look like peanut shells dangling from the honeycomb. The smaller protruding cells (lower left) contain male drones, and the flat, covered cells on the right contain female worker bees.

Grandpa inspecting his backyard hives in 2012, as usual without gloves. He said he didn't mind the occasional sting, and he also believed that bee venom prevented arthritis.

Grandpa drove several beekeeping trucks over his lifetime; this one was his last.
(Photo by Jenn Jackson)

Tending my hives at the San Francisco urban farm, Little City Gardens, 2015. I'd just transferred a new colony into an empty hive and the bees were circling to get their bearings. **(Photo by Jenn Jackson)**

Inspecting a hive for fresh eggs, which tells me the queen is doing her job. The bees store honey in the upper corners of the frame, and the brood nest is in the center. **(Photo by Jenn Jackson)**

The queen, marked with a blue paint dot, encircled by attendants that feed, clean and caress her. (**Photo by MaryEllen Kirkpatrick**)

Grandpa's last beehive, Carmel Valley, 2015.

Wax moth and spider infestation inside Grandpa's last hive. A small colony of wild bees was trying to establish itself in his abandoned equipment, 2015.

A forager returning to the hive with pollen granules packed into concave "pollen baskets" on its hind legs. Pollen serves as protein for the colony.

Hugging my brother, Matthew, after we scattered Grandpa's ashes into the ocean from the Grimes Ranch in Big Sur, 2015.
(Photo by Jenn Jackson)

Franklin Peace, 64, in his favorite deck chair, where he liked to sit and watch his bees return home in the evenings, 1990.

honeycomb inside an empty wooden frame, they suspend themselves from the top bar and hang together like a cluster of grapes to generate heat. When the temperature rises high enough, eight snow-white wax scales emerge from pockets on the underside of their abdomens. One of the bees will emerge from the pack, crawling over all the rest to reach the top of the wooden frame, where she will bite, bend and chew the flakes, mixing it with her saliva until she's satisfied with the consistency. The bee will attach the small blob to the top of the frame and leave. Bee after bee will do this, until there's a small block of formless wax the right thickness to be sculpted into honeycomb.

Next come the builder bees, Grandpa said. They scoop and pull at the wax, taking turns carving hexagon cells. The first honeycomb cell they make sets the mathematical pattern for the rest of the honeycomb, he said.

"Cool," Matthew said, holding his candle up so he could watch the hot wax slide down it and drip back into the can. My nerves settled down with the slow, repetitive movements of candle making, but still I couldn't push the bowling alley all the way out of my mind.

"Grandpa?"

"Mmmmm."

"We got kicked out of the bowling alley."

"Mom got in trouble," Matthew said.

We told Grandpa everything that had happened while he held his candle aloft, forgetting to dip it, and it turned from white to a mustard yellow as it cooled. I saw the muscles in Grandpa's jaw tense as he listened. He set his candle down on an empty hive box and leaned toward us.

"Your mother isn't going to change, so it's best not to upset her. Stay out of her way, and be patient. One day when you're older you'll be able to live on your own."

I told him it was hard to avoid her when we shared a bed.

"Just do what she says and don't talk back to her. Hear me?" He waited for our answer, to make sure we were listening to instructions. We promised to do as he said.

But I didn't tell him that I was afraid of her. It now seemed possible that Mom might actually hurt us.

When the candles were finished, Grandpa cut away the dipping strings, handed us each a pair and told us to bring them to Granny for the dinner table. The delicate yellow tapers were still warm, and smelled of honey butter on fresh biscuits. Granny inhaled their scent and her eyes fluttered. She asked me to fetch her silver candlesticks from the credenza, and then showed me how to carefully polish the heirlooms with a purple paste until they shone.

That night she put one candle on Mom's dinner tray, and three on the dining room table. As Mom dined alone, the four of us ate by candlelight, the flames casting a festive glow over the room as Granny discussed politics, explaining to Grandpa why he, and every other American in their right mind, should vote for Jimmy Carter.

I furtively rolled my eyes at Matthew sitting opposite me, and he giggled conspiratorially. Then he reached his right foot under the table and found my left. We pressed the soles of our sneakers together and pushed our legs back and forth in a seesaw motion, our version of a secret handshake.

We grinned at each other through the pretty candles we'd just made, and for a fleeting moment the day was forgotten.

12

Social Insect

1982

Staying out of Mom's way became considerably easier once I started middle school. I slipped out of our bed an hour earlier now, and walked to my old elementary school to catch a yellow school bus for a half-hour drive to Carmel. Bus seating was based on a pecking order that had been handed down over generations: the eighth-graders in back sitting lengthwise to command entire two-person seats all for themselves, the seventh-graders sprinkled in the middle always politicking for a seat upgrade, and the sixth-graders forced to sit near the crabby driver where he could mean-mug us in the rearview mirror for misbehaving.

But the hierarchy faded once the bus pulled up to the Carmel Middle School campus, which pulsed with several hundred students from all over the Monterey Peninsula. Suddenly I was moving between five different classrooms a day, each with a different mix of people from Carmel and

Pebble Beach and Big Sur. This made me gloriously anonymous. Nobody had to know I was the girl who couldn't listen to the Beatles without crying, or the one whose family was too weird to get her a proper Halloween costume. I blended into the mosaic of everybody, perfectly happy to be one little tile on the wall.

Granny chose my electives, enrolling me in typing and German, and to my great delight, home economics, where I learned to cook and use a sewing machine. The class was entirely female, but I didn't consider it wife-training; I saw it as planning for the adulthood Grandpa promised was coming, when I'd finally cook my own meals without burning them, and never again would I have to wear other people's cast-off clothes.

When a new after-school computer class started, Granny bought a thin floppy disk about the size of a potholder so I could learn to program a machine called an IBM. When the director of the school yearbook asked for volunteers to work on weekends helping him cut and paste all the student portraits onto production pages, I threw my hand up. Whatever my new school offered, I wanted it. I was surprised and delighted that there was so much going on outside our house, and I wanted to try all of it.

Middle school felt like a do-over to my life that had started on the wrong foot, and for the first few weeks I studied people, looking for potential new friends. There was one girl in my English class who held my attention and squeezed. Sophia had the kind of beauty that hushed a room; she was lithe and graceful and looked a bit like Brooke Shields in her Calvin Klein jeans. She carried herself with the cool indifference of a European exchange student who has seen more of the globe than her teachers.

She picked up German faster than anyone in class, and when she sat next to me in English, her long, dark hair swished when she flipped it out of her face. She smirked a lot, and I desperately wanted to know what she was thinking, what kind of music she listened to and where she went after school. She told me that she was allowed to drink red wine with dinner, and that her mother sometimes took the passenger seat and let her drive their stick-shift LeCar to school. I didn't doubt it. Sophia was so beguiling that high school boys were already dedicating love songs to her on KSPB, the radio station at the private school in Pebble Beach. During written tests, whenever she leaned toward me to whisper that she didn't know an answer, I turned my paper so she could copy mine. I didn't care if I got caught.

One day I got up the nerve to ask her what shampoo she used to make her hair smell so good.

"Something from my mom's salon," she said.

The word *salon* lit up like the Hollywood sign in my mind. I was still going to the village barber and getting a lollipop after he chopped my hair into the same helmet cut. I muttered something about how great it must be to get expensive shampoo for free, then immediately regretted sounding so unsophisticated.

"I can get you some," she offered. "Come with me to my mom's salon after school. She won't mind."

Imaginary game show glitter shimmered down all around me.

"You sure?" I said, doing my best to appear undecided.

The rest of the day was a blur, and after the last school bell, I met Sophia behind the gym. She led me on a shortcut through an open field that emptied fifteen minutes later at The Barnyard, a boutique shopping center designed to look like a cluster of barns surrounding a big windmill. It was a touristy place where visitors bought cashmere sweaters or oil paintings of the Central Coast, but its flagship store was geared more to the locals: a massive bookstore with an organic café in back. Sophia led me along The Barnyard's brick garden paths, up a flight of stairs and down a balcony. I heard the whine of hair dryers and knew we were close. Sophia pushed open the door, and an Adam Ant dance song came thumping out, all trumpets and drums.

"Honey, is that you?" called a voice from behind a screened partition. "I'll be out in a sec."

Sophia slid into one of the architectural leather and chrome chairs in the waiting area, drooped one leg over the arm and flipped through a *Vogue* magazine. She studied each outfit with laser focus, absentmindedly licking her fingertips before flicking each page. It now made sense why Sophia glided through campus on her own invisible runway; she was homeschooled in fashion. I heard a faucet shut off, and Adam Ant faded from concert volume to background music.

Sophia's mom walked into the room, and suddenly I felt as if I had stumbled right into an MTV music video. She was the spitting image of Pat Benatar, an elfin beauty with spiky black hair, high cheekbones and stage-ready makeup. She shimmered in a gold jumpsuit with shoulder pads, one forearm sheathed in sparkly bangles, her pixie body perched on stiletto-heel boots. She'd accentuated her eyes with heavy kohl and more mascara than I knew eyelashes could hold. Her eyelids were metallic purple that blended to neon blue toward her brows, and the only missing accessory was an electric guitar. She scooped Sophia into a hug and kissed her on both cheeks as if it had been years, not hours, since they were last together.

Then Sophia introduced me, and Dominique leaned in to leave lipstick prints on my face, too. Like a weak plant that had finally been moved into the sun, I lifted my cheek to meet her halfway.

"Enchantez," she purred.

"Means nice to meet you," Sophia said.

"On shon tay," I repeated, too starstruck to find my own words.

Dominique and Sophia chattered and giggled like two best friends at a coffee shop, swapping stories about their day and finishing one another's sentences. Dominique made jokes about a rude customer, and Sophia told her mom that our English teacher was on her Don Quixote kick again, making the class rehearse lines from the play even though there were no plans for a school performance. I studied them with a mixture of wonder and longing.

Dominique asked me how I liked school, and I told her I loved everything about it except for dodgeball. Whenever it rained, the PE classes were stuck inside the gym where the teacher separated us into two teams, gave us a rubber ball and ordered us to hurl it at each other. I cowered in the back rows, hoping I'd somehow get through the ordeal without too many bruises. Bullies were the only kids who appreciated dodgeball.

"So barbaric," Dominique said, reaching her fingers into my hair to feel the texture. "Needs moisture," she said, and led me to the shelves of bottles. Dominique pulled down three potions and unscrewed the caps, letting me smell them. I chose the one that smelled like tangerines.

"Good choice," Dominique said. She put it in a small gift bag with handles, and it looked like a birthday present.

Sophia and I spread our homework on the coffee table in the waiting area. Dominique set a bottle of Pellegrino before us and handed Sophia some cash, and told us to get sandwiches so we could eat while we studied. I practically floated on the brick path to the bookstore-café. Sophia had my Fantasy Mom.

After her last customer, Dominique drove us home in the yellow LeCar. I sat in the back while Sophia rode in the passenger seat and reached over to shift gears when her mother pressed the clutch and called out a number. Sophia was so practiced that she didn't even have to look at the diagram on the shifter when her mother accelerated, proving she had been telling the truth when she said she already knew how to drive. On the way to the Valley, I learned that they lived just a few miles from me, and Sophia had an older sister. The three of them lived together, also without a dad at home. Yet whatever had split their family apart didn't freeze them in tragedy. Dominique was still Sophia's mother, all the way.

When Dominique asked about my family, I abbreviated and said that I lived with my grandparents. I relaxed when neither Dominique nor Sophia asked for an explanation. I directed Dominique to our driveway, and as I was getting out, she pointed at the honey bus.

"What's that?"

"Grandpa's honey bus."

"Honey bus?"

"He makes honey in there."

"He's a beekeeper?"

They had a zillion more questions. They wanted to know where his beehives were, how bees make honey, how we get it out of the hives and how many times I've been stung. I led an impromptu beekeeping 101 lesson, describing the hive as a superorganism with one collective brain.

There's a queen, but no king, I explained, but the queen isn't the ruler. All the bees work and make decisions together. Bees are loyal and generous, but they also have a brutal side and will toss out the weak, the sick and the males once they become useless to the community. Bees have their own language, and will hum with delight, shriek in distress, fall silent with grief, and growl with menace when threatened. Even the queen has her own special battle cry when rivals challenge her throne.

I basked in the attention of my audience, growing more confident as I tried to talk to them the way Grandpa spoke to me, in stories with a little bit of pizzazz layered on. I made them guess how many eyes bees have (five), then added that their hairy eyeballs can see ultraviolet light, picking up psychedelic flower colors and patterns we can't see. Dominique asked if it was dangerous to open beehives.

Beekeepers can't be scared, I said ominously, because bees can smell your fear.

Dominique and Sophia exchanged a look.

"Grandpa says it's true," I added.

Bees don't like bad breath, either, I continued. Or dark colors, so beekeepers brush their teeth and wear white suits so the bees don't mistake them for bears. Sophia and her mother hung on to my words, so I even told them the gross stuff—that bees have sex in the air and the male bee dies after mating because his *thingy* breaks off inside the queen. I told them honey is actually nectar that bees have barfed up and fanned with their wings until it thickens. I was really working it, trying to wow them.

When I finished, it was silent in the car for a moment. I think they were deciding whether I had an active imagination.

"That's so…*cool*," Sophia said.

It felt backward, her admiring me, but it was the best kind of wrong I'd ever felt. I knew now, without a doubt, that we would be friends. We both had a currency the other wanted.

"Don't forget this," Dominique said, handing me the shampoo I'd forgotten on the back seat. "Come back again anytime."

"Tomorrow?" Sophia asked.

Yes, yes indeed.

We fell into stride after that, walking to the salon together several times a week. I ate dinner with Sophia's family so often that I was like their foreign exchange student, such a part of the routine that Dominique put out a toothbrush for me and Sophia gave me her old Gloria Vanderbilt jeans and Lacoste sweaters.

Granny granted me permission to spend time at Sophia's house, and I lived vicariously through my chic new friend. She and her mother took me to French restaurants and introduced me to escargot and my first taste of red wine, and they took me to see *Fast Times at Ridgemont High*, my first R-rated movie. Even though we were the same age, Sophia seemed more like a grown-up to me. Her bedroom was filled with cubist Scandinavian furniture that we were forever stacking into different configurations. We blasted her stereo and redecorated for hours, stopping every now and then when a boy called her private telephone line. I sat nearby and pretended not to listen to her flirt, but in truth was taking meticulous notes so that if anyone ever fell in love with me, I'd know what to say. Sophia liked to stay up late watching movies, and on the nights I fell asleep at her house, her mother took us both to school in the morning.

I couldn't envy Sophia because she treated me like a sister. But the more time I spent with her family, the harder it became to return to mine. Mom's absence was amplified now that I'd spent time in a home filled with laughter, dinner parties and music. In Sophia's house, single motherhood was not the defeat that it was in ours. Dominique was built of something stronger than my mother, and that made me increasingly impatient with Mom because it seemed like she wasn't really trying. What was the statute of limitations on sadness?

As Sophia and her mother let me borrow more and more of their happiness, I felt increasingly selfish because I could never reciprocate. A few times Sophia asked if she could come to my house, but I always deflected, saying vaguely that my mother was sick. I felt ashamed of my mother's weakness, and didn't know how to explain that she had retreated behind a closed door. My life seemed so flat compared to Sophia's, and I was afraid to let her see how much sharing my family had to do, of beds, of bathrooms, of grief. I didn't think Sophia would understand, and I wasn't sure if I could have explained it in a way that would have made sense to her, anyway.

I spent less time sharing the bed with Mom, which I think suited both of us. She didn't ask where I was, so I didn't mention Sophia, assuming Granny must have filled her in. More and more, I began to feel like I was leading a double life.

Early one Saturday morning, I awoke to the smell of hazelnut coffee and found Mom at the kitchen table warming her hands on a steaming mug with the *Monterey Herald* spread before her. She never followed the news, so I peered over her shoulder to see what she was reading. She was circling garage sales, selecting the ones in the high-end neighborhoods. She blew out a column of smoke and looked up at me.

"If we leave now, we can get there before all the good stuff is sold," she said.

"We?"

"Got anything better to do?"

I wasn't sure if this was the best idea. My last outing with Mom had ended with a near arrest in a bowling alley. She fished the keys out of her purse.

"C'mon. I'll let you pick out one thing."

Sold. I couldn't resist a free gift.

The Gremlin complained as Mom forced it up the snaking, two-lane Los Laureles Grade in a high gear. She slowed before mailboxes to check the house numbers, until she found the one listed in the paper, and wound the car through a pillared gate toward a home as big as a hotel, with a view of the entire valley. I spotted a tennis court and the turquoise of a swimming pool through a slatted fence. We parked near a fountain with a bubbling stream coming from a fish's mouth and walked to the garage, where a woman was pulling books out of a box and setting them up on a folding table. We were an hour early.

"Oh, you're…the first ones here," she said, pulling back her cuff to check the time.

"That's great!" Mom said. "Then you can show me the good stuff."

The woman forced a smile and brought Mom to a table with crystal vases and china plates.

"It was my aunt's wedding set," the woman said.

Mom slowly perused the wares, turned over each item to inspect the price tag, and then gently placed each thing back down again.

"What a rip!" Mom whispered to me too loudly, and I cringed, hoping the woman hadn't heard. Mom walked the perimeter of the garage, handling everything as if she were looking for clues. She held sweaters up to her chest and checked the arm length. She flipped through books. She even examined things I know she had no intention of buying, like a power drill and a set of skis.

I tried to fade into the wall, watching the homeowner

watch her. We were trying to figure out what, exactly, she was doing. Then, it dawned on me. She wasn't shopping at all; she was only here because she enjoyed sifting through other people's lives. Rich people's lives.

I tugged on her sleeve. "Can we go?"

"We'll go when I *say* it's time to go," she hissed under her breath. Then she turned from me to the homeowner, her face soft and kind.

"Excuse me, would you mind if I used your bathroom? I'm so sorry to have to ask." She lowered her voice and whispered confidentially, "It's medical."

The homeowner looked startled. She hesitated, and then asked Mom to be quick—she couldn't leave her yard sale unattended. The woman let us in and down a hallway with windows in the ceiling that let in columns of light that made bright yellow squares on the terra-cotta floor. Mom followed slowly so she could catalog her surroundings. She ran her finger along a glistening countertop, took note of a refrigerator that poured ice and water right from the door, and quickly glanced inside rooms. I tagged behind, humiliated that Mom would stoop to such bull honky to get inside the house. The woman showed her to a bathroom, and Mom clicked the door behind her. I could hear her opening cupboards and medicine cabinets, looking for clues to what her life might have been had it gone the way it was supposed to. The homeowner and I stood next to one another, awkwardly clearing our throats and listening to Mom rummaging.

"You okay in there?" the lady said, rapping on the door.

I heard footsteps, then a flush, then Mom ran the tap for a second and whipped open the door.

"Oh, hi!" Mom said cheerfully. "I just love that spa tub you have in there."

The woman gave a wan smile, and there was an uncomfortable silence. "Well, we really should get back outside."

We fell in line behind our reluctant tour guide, but Mom wasn't ready to give up so easily. She kept up her squirrel chatter behind the woman's back.

"Who is your contractor? It's so hard to get a good one these days. My husband wants to remodel the bathroom with a spa tub, and at first I was against it because we already have an outdoor hot tub. One of those redwood ones, you know? But looking at yours, now I'm starting to change my mind again. Do you use it a lot?"

The woman didn't answer, and once we were outside, she strode away from us and rushed up to a man I assumed was her husband, because he immediately glared in our direction. I was mortified that Mom had crossed a line, got caught and wasn't even aware of it. She had stolen something from these people, even if it wasn't an actual thing you could hold in your hand. She had taken a little piece of their privacy for herself. I was ashamed, and I needed to get Mom back in the car before she caused any more damage.

"Mom, we should go."

She opened her mouth as if to protest, then caught the man looking our way. She linked her arm with mine and leaned in as if to tell me something in private, but projected her voice. "Nothin' but a buncha overpriced crap here, anyway."

I tugged her toward the car and picked up the pace.

"What's with you?" she said.

"I'm just cold."

Mom had several more sales on her list, but I talked her into taking me home, saying that Grandpa was waiting for me to check the bees. It wasn't exactly true, but I could easily make it true once I got home and found him tinkering around in the yard. All I had to do was tell him I felt like seeing the bees, and he'd drop whatever tool was in his hand and pick up a veil. I just needed to get Mom home within the safety of four walls again, where she couldn't embarrass me, or get in a fight with someone.

Ever since we'd arrived in California, I'd carried a constant hope that she'd return to society. But the few times she'd left the house, it seemed that nothing ever went right for her. She had a knack for getting herself removed from every place we tried to go, and her self-righteousness afterward always embarrassed me. Her anger was capricious but always guaranteed; she flared at the slightest thing—a driver who forgot to signal, a grocery checker who refused an expired coupon.

As my perimeter expanded beyond Via Contenta, I was increasingly beginning to suspect that Mom's unpredictable moods were a part of her personality, not just a temporary sadness over losing Dad or her difficult lot in life. All that bed rest hadn't improved her outlook. She moved through the world defensively, assuming the worst in others, and she was even more convinced that people were out to get her. I worried that if I upset her, she could just as easily turn against me. Sometimes I even thought it would be safest if she stayed in bed permanently.

I sought refuge in the bee yards, and the more time I spent with Grandpa, the more I began to appreciate how easy it was to be with him. We could talk, or not talk; it didn't matter. We enjoyed each other's company, and the simple ease of that made me feel that maybe things weren't so bad after all. I wondered what it was about Grandpa that didn't come naturally to me and Mom. As I became more curious about this, I started to wonder who he was before I showed up on his doorstep. Who taught him all the things he was teaching me? It dawned on me that Grandpa must have been someone else before he became my grandpa. But I knew very little about this man who had become the most special person in my life.

On one of our drives to Big Sur, I finally asked him why he was a beekeeper.

"Well, my dad kept bees, and his father kept bees, and my cousins kept bees. There were beehives on the Post Ranch, where my mother was born. Her daddy and granddad keep bees, so I guess I just did, too."

"Why do *you* like it?"

We slowed to a stop on Highway 1 as an RV in front of us lumbered toward one of the oceanside pullouts, where tourists were snapping photos of the single-arch Bixby Bridge that connected two sections of coastline. Grandpa waited patiently with the truck idling.

"Well...you can work by yourself. People don't bother you. You have to move slowly when you work the bees, so it's a job that's calm, I suppose. And people always like it when I give them honey."

The RV was out of our way now, and Grandpa exchanged a wave with the driver as we continued south.

"And Big Sur is a good place for bees," Grandpa continued.

"Why?"

"I have to take good care of them and put them in a place where they can fly free."

I was confused. Can't bees fly wherever they wanted?

He unscrewed his thermos while keeping one hand on the steering wheel, and handed the cup to me, signaling me to fill it with coffee. I waited for him to let the caffeine kick in, and then he rolled down his window and rested his elbow on the door, settling in to explain something to me.

"There are three different kinds of beekeepers," he began.

Hobbyists, he said, keep a handful of hives to learn about bees and harvest a little honey; sideliners like himself run small businesses with more than a hundred hives in fixed locations; and then there are the big guys with thousands of hives who truck their bees across the country to pollinate huge agricultural farms.

"Those migratory beekeepers don't even bother with honey. They make all their money renting bees to farmers," he said.

I had never imagined beekeeping any other way than how Grandpa did it. He worked in harmony with the bees, attuned to their needs. It was hard to believe that outside Big Sur it was the other way around. Bees were shuttled on the highways and forced to work for humans.

"Where are all those bees going?"

Mostly to the almond farms in the Central Valley, he said. There aren't enough bees in the whole state to pollinate all the almond flowers, and the trees depend on bees because their pollen is too heavy for the wind to carry. Beekeepers come in from other states, and use forklifts to lower their hives into the orchards, leaving the bees there for several weeks in spring to pollinate rows and rows of almond trees as far as the eye can see. Bees need a diet of diverse pollen to stay healthy, he said, but traveling bees are forced to eat the same thing, day in and out.

"Imagine eating a hot dog every day for a month; then a hamburger every day for a month," Grandpa said. "What do you think would happen to you?"

"I'd probably throw up," I said.

"Exactly."

Once the bees finish pollinating one farm, the beekeepers retrieve their hives and haul them to the next crop in bloom, unleashing their bees on cherry farms in Stockton or apple orchards in Washington. Bees-for-hire toil from February to August, which means that a typical honeybee in America spends more time on a highway than in the wild.

"That's why I don't move my bees," Grandpa said. "I think those commercial bees are stressed out. It's not natural to take bees out of their environment. They get disoriented, and it takes them a while to establish themselves again. It's too hard on their system."

It's not the traveling alone that does the bees in, Grandpa said. It's also the crop pesticides the bees pick up and absorb into the architecture of their hives. Like living in a home with lead-based paint, the exposure can be undetectable at first, but over time the bees develop nervous system disorders, lose the ability to fly and die.

"That's why I put my bees in a place far away from people, where there are no chemicals. So I can protect them."

Grandpa's bees were safe, but now I was worried for the traveling bees. Were they all going to get sick and die?

"Are bees in trouble?"

"Not yet," Grandpa said. "But if we keep treating them like slaves, we could lose bees for good."

"Then what?"

"Then we don't eat."

There was the answer to my question. Grandpa was a beekeeper because he understood the things that really mattered.

He knew that there should be a balance between the taking and the giving a person does in one lifetime. That a good relationship, between bees and humans, or two middle school classmates, or between a mother and daughter, all needs to start from a mutual understanding that the other is precious.

13

Hot Water

1982

Not long after I started middle school, there was an abrupt change to my family's living arrangements. The rental house next door became available, and Granny seized her opportunity. No sooner had the neighbor lady, who made wicker bassinets, packed up her last bit of thread than Granny had an announcement: Mom, me and Matthew were moving in and she would pay the rent. As part of the deal, Mom would need to get a job to pay for utilities and groceries. The carrot worked. Mom found part-time work as a loan officer in a bank. Finally, seven years after we arrived, Granny was getting her house back.

Our new home was even tinier than Granny and Grandpa's, had no shower or heat, and the floorboards were warped in places, but it was all ours. The linoleum in both the bathroom and kitchen was chipped and cracked, the screen door listed to one side, and there were cigarette burns in the moss green carpet, but none of it mattered because I believed this little rundown house was where we were going to finally become a family again. Out from under Granny's wing, Mom could start over as our parent. This house would be our comeback, and maybe, just maybe, one day when things were all better again I could invite Sophia over.

There were two bedrooms at opposite ends of the house. Mom took one, and Matthew and I would share the other one, which had been converted from a garage. Our bedroom door opened to three descending steps, and the floor was concrete and covered with a thin tan carpet pad instead of an actual carpet. There were two windows, at waist level, on opposite walls. The room was cold, and the rough-hewn knotted pine walls lacked insulation, but its saving grace was two closets, giving Matthew and me our first bits of personal space.

Granny went to the auction house in Monterey for our furniture. She bought a set of bunk beds and a vintage Western dresser with an age-spotted mirror for my brother and me to share. The auction house delivered a twin bed, a laminated wood dresser and a one-drawer side table for Mom's room. Couches were too pricey, so Granny bought a scratchy floral print love seat with wooden armrests instead. It was the only seating in our living room, and the most impractical choice for three because only two people could sit on it at once. We became the new owners of a flimsy bookcase that wobbled when you took a book out, and a six-inch black-and-white television with rabbit ears that Mom placed on the mantel over the fireplace, where it was too far to see from the love seat. The final touch was a portable record player that she set on top of the tiled coffee table, so she could play her three albums in rotation: *Saturday Night Fever, Grease* and *The Bee Gees.* Mom decorated with macramé plant holders and spider ferns she picked up at garage sales.

On move-in day, Matthew and I carefully arranged and rearranged our clothes and shoes in our two closets.

"Hey," Matthew said. He poked his head out of his closet, where he was stacking his Lego sets on the shelf.

"What."

"Is there anything to eat in the kitchen?"

"Go look."

"No, you."

"You're such a baby," I snapped.

The refrigerator was an avocado color, to match the oven. I winged open the door and found little on offer: a six-pack of Fresca, an enormous tub of low-fat cottage cheese, celery sticks, half a shriveled grapefruit and a package of English muffins. Mom was dieting again. I opened all the cupboards until I spotted a bowl, and spooned some cottage cheese in.

"What do you think you're doing?"

I jumped back, feeling suddenly guilty.

Mom whisked the bowl from the counter and tipped the contents back into the cottage cheese container, resealed the lid and returned it to the fridge. She slammed the door for emphasis.

"First of all, that's food I bought for me. You can't just take things around here," she said. "Second of all, don't leave the refrigerator door open like that. You'll let all the cold air out."

And thus the new order was established: the house was hers, and Matthew and I just happened to be living in one small corner of it. I washed the empty bowl and tried to remember Grandpa's advice to not let Mom upset me. I wanted to protest but knew it was futile. When Mom got mad, she was like a train in motion that couldn't be pulled off the tracks. She had that prickly kind of anger that wanted to attach itself to everyone around her, as if she already knew she was going to be mad for the rest of her life and wanted company. I said nothing while I dried the bowl and returned it to the cupboard, leaving her waiting for an apology as I returned to my new bedroom. It had been naive to think Mom would find Matthew and me less irritating simply because we switched locations. A person's beliefs don't change with the scenery. I let my hope go just as easily as it came, a pretty ribbon fluttering out of my hand. Matthew's face fell when he saw me return empty-handed.

"Let's go check Granny's fridge," I said.

Over the next few weeks, we learned Mom's house rules. The food was carefully separated into diet sodas and low-sugar snacks for her, and microwave meals for us to heat up on our own: frozen burritos, hamburgers and TV dinners. But her possessiveness extended far beyond groceries. My brother and I needed permission to turn on the television, use the phone, or plug in the space heater. Now that she had bills to consider, Mom calculated every watt and every drop of water we used. When one of us took a bath, Mom listened outside the door and banged on it when we hit her perceived water limit. My brother and I learned to use the appliances before she got home from work, and to turn the television off at least an hour before her arrival so it would cool off and not give us away. She retaliated by dragging the TV into her bedroom so we couldn't use it. Then she dragged the phone in. Then the radio, until eventually we saw less of her than we had at the old house. It didn't take Matthew and me long to migrate back to Granny and Grandpa's home for warm meals, uninterrupted showers and television.

When she couldn't cover the bills, Mom started migrating next door, too. First, to save money she canceled her garbage service and started putting her trash bags in Granny and Grandpa's cans. She did her laundry at Granny's to save on water. Then she came over to borrow milk, or butter, or filch some wood from Grandpa's woodpile. Granny started giving her a monthly allowance so she'd stay in her own house.

My favorite place in the new rental house was the bathroom because it afforded the one spot of true privacy. I liked to disappear for a good hour in the tub reading one of my Hardy Boys mystery books until the water went cold. I was doing exactly this one afternoon when I got the idea that I could extend my bath and read longer if I let some of the tepid water out and reheated with more hot. I knew it was risky because Mom might hear that I was using more than a tubful of water, but if I just pried up the bath plug a millimeter with my toe and let the water trickle out quietly, maybe she wouldn't hear it. It took a long time, but eventually I'd let out half the water. I turned the hot faucet slightly, and put the washcloth under the stream to muffle the sound. The warmth of my glorious insurrection pooled around my legs, and when steam rose again from the water, I relaxed back with my book once more.

Two sentences in, I heard footsteps gathering speed, and the bathroom door burst open with a bang. Mom wrenched off the tap, grabbed the book out of my hand and chucked it into the wall. She grabbed the side of the tub and leaned toward me so that her hot breath mixed with mine. She seemed feline, like she was leaning in to smell my fear.

"What the *hell* do you think you're doing?"

I tried not to make any sudden movements. She and I both knew exactly what I had been doing. Stealing water. Mom grabbed my upper arm and yanked me out of the tub so fast that I had to cling to her to keep from stumbling. I found my footing and stood there dripping as she blocked the exit with her body. She was seething, her face a shade of red I'd never seen.

"Don't think you're smarter than me," she said, jabbing her finger at me.

"I don't."

I was starting to shiver. I needed to figure out how I could get around her and out the door. Maybe if I apologized.

"I'm sick and tired of both you kids wasting water. You seem to think I'm made of money. Well, listen and listen good—I'm not."

"I'm sorry," I muttered.

Truth was, I wasn't sorry at all. I was mad as a teapot. Water consumption never came up in Sophia's house. If we needed to wash dishes or take a shower or flush the toilet, we didn't think twice. But at home I fretted over water all the time, and just the sight of it made my stomach knot up worrying about its preciousness. I knew I shouldn't have tried to take more than my share. My mind scrabbled around trying to figure out how to calm her down and get a towel.

"You don't sound sorry."

"Can I have a towel?"

Mom narrowed her eyes. "I'm not finished with you."

I didn't know if she had just given me a reprieve, or a threat. But I didn't wait to find out. I lunged for the towel rack and ripped down a towel, then scooted behind her and out the door before she had time to react. I ran for the bedroom hoping Matthew was there, because two of us stood a better chance against one.

Before my mind had time to register what was happening, I felt her weight land like a mattress on my back. I pitched forward and landed on the carpet with such force that the wind knocked out of me. I felt time stop as I searched for my breath again, and then felt myself being rolled like a rag doll onto my back, and then Mom pinning me underneath her like a wrestler. Her body pressed on me like a bag of sand, and I gasped for air.

"All you kids do is take, take, take! After all I've done for you! I've had to do this all on my own, but do I ever get a thank you? Noooooooooo!"

My heart thumped against her inner thigh as I slapped at her arms and tried to buck, but I was trapped. Adrenaline coursed through me, and I thrashed as hard as I could, but I couldn't budge her. We were like two cats pawing at one another as she tried to grasp my flailing arms. Finally she caught my wrists and wrenched my arms to my chest, where she held them folded. Her lips tightened in fury and she shouted over me, at a spot on the wall.

"You have no idea the hell I lived with!" Her nonsensical outburst shocked me into submission, and I stopped struggling, uncertain of what was happening. She seemed to be talking to someone I couldn't see.

"Nobody liked me. Nobody EVER liked me!"

A quiet terror filled my lungs. Mom was somewhere else in her mind, in an altered state where I couldn't reach her. The voice that came out of her was familiar, but much younger, how I imagined she sounded as a small girl. It seemed possible that she wasn't even aware of what she was doing. And this was most frightening of all, because what if she was capable of doing far, far worse to me? I begged for release, but my words bounced off her, unheard. Her anguish distilled into a one-word drumbeat.

"Nobody! Nobody! Nobody!"

She buried her hands in my wet hair, curled her fingers around two hanks of it and pulled. An instant, white pain of a thousand needles pricked my scalp. She yanked my head side to side, and we were both screeching now, unintelligible sounds like trapped animals wailing for rescue. I felt my follicles rip away, and from the corner of my eye I saw my hair slip from her fingers and flutter to the ground. I squirmed to get loose, but she shifted her weight slightly to block my escape. I had no way out.

I went limp, giving in to whatever was coming next. I closed my eyes and saw myself sinking toward the bottom of a dark ocean, floating farther and farther away from her. As I descended, it got quieter and quieter, until her screams dissolved. I gently floated down to the bottom without sight, without sound. As I rested on the soft sand, retracting steel doors slammed down around all four sides of my heart, boxing it in where she would never reach it again.

That's when I made up my mind that I no longer belonged to her. As soon as the thought came to me, a warm light broke through the darkness all the way to the seafloor, warming my skin all over. I was free. She could do whatever she wanted to me now, and it wouldn't matter. I was mine now, and would never again be hers. Relief enclosed me in a cocoon, knowing that I didn't have to love her simply because she was my mother. All I had to do was survive her, and one day I could leave her for good. Grandpa was right. If I just obeyed and kept out of her way, I'd survive. My body was imprisoned beneath her, but my mind didn't have to be. The thought made me smile.

"Oh, you think this is funny?"

She raised her palm and her slap was quick and sharp, sending an electric jolt across my cheek. I tried to cover my face with my hands and turn my head away, and through my fingers I spotted Matthew coming out of the bedroom just as Mom swiped her fingernails across my opposite cheek.

"Mom!" he shouted. "Stop hitting her!"

His voice landed on her like a lasso and she instantly stilled. She looked down at me with a quizzical expression, as if she didn't recognize me. She gasped once, rolled off me and slumped on the carpet, her shoulders heaving. I scuttled like a crab in the opposite direction, backing up to the wall so I could keep my eyes on her. She was sobbing now, rocking back and forth with her arms wrapped around her knees. I reached up and touched my hairline, and pressed on the bald spot to make it stop throbbing. I rose to my feet and crept on wobbly legs along the wall to the bedroom, and quickly pulled on some clothes. I heard the bedroom door creak on its hinges, and froze.

"It's me," Matthew said, sticking his head inside the room.

He came into the room and reached for my hand, and we ran past our balled-up mother, out of the house, through the fence and to our grandparents' house. Granny and Grandpa were watching TV when we thundered into the living room, talking over each other in a hysterical rush of words.

"Whoa, slow down," Granny said. "One at a time."

I tried to explain but sputtered halfway into sobs, so Matthew finished for me, telling Granny what he'd seen. Grandpa reached for the lever on his recliner and bolted himself upright. Granny scowled and snapped off the television. "Well, what did you do to upset her?"

"Ruth honey!" Grandpa said, shooting her a pleading look that did absolutely no good. He had corrected her, and she was incredulous.

"I beg your pardon?" she said to Grandpa, like she was berating one of her insolent students.

Grandpa turned to me. "Are you hurt?"

"She doesn't look that hurt to me," Granny said, squinting at me from across the room. She turned for the bedroom, complaining to an invisible audience as she walked away. "If it's not one thing, it's another. I swear to Almighty God I'm going to get some peace one day before I meet my Maker."

I heard the clacking of the rotary dial as she called Mom, followed by murmurs of consolation. It was going to be Mom's word against mine.

Grandpa shook his head in disgust, and I thought he might make a complaint, but he held on to whatever was on his mind. He stood and let out his breath, like he had been holding it awhile.

"Let's go on outside," he said.

Without needing to discuss it first, the three of us walked toward Grandpa's beehives. There was more activity than usual outside them, and at first I thought one of the colonies might be swarming. But as we got closer, I could see it was only a group of bees circling outside the hive. They took to the air, made a small loop in front of the hive and then returned to the landing board. They repeated the pattern over and over, as if they kept losing their courage to travel.

"What are they doing?" Matthew asked.

"Practicing," Grandpa said, handing me a hive tool and the smoker to Matthew. Grandpa and I lifted the lid off the first hive while Matthew smoked the entrance.

"Practicing what?" I asked.

When a house bee grows up and is ready to start gathering nectar, it doesn't just one day zip out of the hive ready to go, Grandpa explained. It has to learn to fly first.

"Every day at about this time, the bees have a flying class. They make lazy eights in front of the hive, memorizing landmarks and the angle of the sun, so they will be able to find their way home. Each day they make bigger and bigger loops, following the older bees until they are steady on their wings. They don't go to the flowers until they feel ready."

"How long does it take them to learn?" I asked.

"I dunno. It all depends on the individual bee, wouldn't you say?"

It made sense. I didn't just one day race out of the house and know how to read, or do math. I had to go to elementary school down the street and practice. Then when I got older and more confident, I traveled farther on the bus to middle school and studied some more. Soon my circle would widen again when I'd start high school. Like the bees, I learned by trying and failing, over and over again until I got it right.

He lifted out a frame and tilted it in the sunlight, checking the comb for eggs. I watched the bees repairing cracks in the wax, grooming one another with their forelegs and mandibles, and dipping their heads into the brood cells to feed the larvae. Everything was as it should be inside the hive. I could rely on the bees to always be working, each with a purpose and a rhythm that soothed me. I felt the knot of apprehension in my stomach unclench and my shoulders relax.

Grandpa held a honeycomb frame in front of his face, and spoke to us from the other side.

"Do you feel like talking about your mom?" he asked.

My brother and I looked at each other, each waiting for the other to go first.

"I don't want to go back there," I said.

"You two can stay here tonight," Grandpa said. "Don't worry, we'll figure something out."

My brother plucked green grass and stuffed it in the spout of the smoker, and handed it back to Grandpa.

"What set her off?" Grandpa said.

Matthew looked away to the neighbor's yard, as if the memory was too much for him to repeat.

"I was sneaking hot water."

Grandpa shook his head. "That woman," he muttered.

Just then, Granny's voice floated over to us. She was

standing in the doorway with the phone receiver in her hand, stretched all the way across the kitchen.

"Meredith! Come apologize to your mother."

I flinched. What I'd done was wrong, but how Mom responded was more wrong. I was not going to apologize.

As I lay trapped beneath her, a terrible suffering had poured out of her in a fugue, revealing a brokenness in my mother that completely unnerved me. She had been yelling at someone in her past, but hitting me in real time. This was not something that an apology could even begin to fix. Mom was having serious trouble, yet no one seemed to want to take it seriously.

Seven years had passed since we left Rhode Island, and Mom was still as despondent as the day we arrived, if not more. Each year that her luck didn't change, her downward spiral sped up a little more, making it increasingly difficult for anyone to pull her out of her funk. I had hoped work would give her a diversion, but she clung to her victimhood even harder. She came home from the bank outraged at how rude customers were to her when she couldn't approve their loans. Her boss was incompetent, and her back hurt from being on her feet all day. Her coworkers were lazy imbeciles, and she was always being called in to cover their shifts. Nothing was ever, ever right. The anger was building inside her, in layers, a little bit more each day until eventually it would consume her.

If she attacked me without warning today, odds were good that she could attack me tomorrow, or next month, or next year. Apologizing was tacitly agreeing Mom's aggression was nothing to be concerned about, and that I in some way brought it on myself. I knew better now. Going forward, I vowed to stay as far away from Mom as possible.

Granny repeated her demand, a little louder this time. I looked at Grandpa. I needed him to stand up for me.

"Wait here," Grandpa whispered. "I'll tell her you're too shook up right now."

Grandpa was able to postpone my apology, and Matthew and I went to bed early to avoid the constant volley of phone calls from Mom trying to reach me. As I lay under the covers waiting for sleep to come, I remembered the night Dad had asked me if I'd rather live with him. He had also asked if Mom ever hit me, and I'd been shocked by the suggestion. Had he been trying to warn me? What did he know about Mom that made him think she might do this?

"You awake?" I whispered.

"Yeah," Matthew said.

"Thanks."

Matthew sniffed. I couldn't tell if he was crying. "You'd do the same for me."

"Of course," I said.

"You okay?" he asked.

My cheek was still hot where she'd scratched me.

"I will be."

I slept in fitful snatches, interrupted by worries that I'd picked the wrong parent.

The next morning, I looked into the bathroom mirror and saw evidence of last night's row: four long welts on one cheek where Mom's fingernails had raked from my eye to my chin. They were sore and enflamed, protruding from my face like fat, red worms. I looked ghastly, but there was no way I was going to stay home from school. It was safer there. If anybody asked, I'd say that Matthew and I had been roughhousing. I stuck to my story, but some of my teachers hesitated a beat before deciding to believe my lie.

Matthew and I continued to stay the night in the little red house for the next few days, while Granny kept in nightly phone consultation with Mom. It never made sense to me why they communicated this way, at arm's length, when one of them could have simply walked the twenty steps separating their homes so they could talk in person. I sensed something serious was being negotiated, and at some point apologies were going to be extracted from Mom and me. I assumed there would be consequences, but none ever came.

Instead, Matthew and I got on a plane for our annual summer visit to Rhode Island, and we didn't mention anything to Dad, afraid he'd pull us out of California to a life we didn't know. Mom's outburst became another unmentionable, obscured behind the thick curtain of family history.

While we were away, Granny bought a used camping trailer, and had Grandpa tow it to the house and park it near the honey bus. It was a white aluminum box with a set of rear wheels, about fifteen feet long and fit no more than two people at a time inside. It had windows with horizontal glass partitions that cantilevered outward, a twin bed on one side and a dinette opposite, with a sink, mini-fridge and closet in between. It smelled slightly moldy, had no heat and made absolutely no sense because our family did not go camping.

When we returned from the East Coast, Granny announced that the trailer would be Matthew's new bedroom. We were getting too old to share a bedroom, she said by way of explanation. My brother and I took this information as truth because Granny had said it, but at twelve, and ten, we had never felt burdened by one another in our shared room. I felt ashamed by the implication that my brother and I had been doing something wrong, and didn't understand why getting older could be a bad thing. Instead of the gratitude Granny was expecting, we stared at her blankly, both of us feeling a vague sense of loss.

Matthew and I stepped into the trailer and looked around, testing the firmness of the mattress and opening drawers. He turned the tap, but no water came out because Grandpa hadn't yet hooked up the hose. I instantly became envious. I was the one Mom fought with, why wasn't I the one getting rescued? Now I was alone in the house with her. What if Matthew couldn't hear me the next time I screamed? My brother saw my long face and tried to cheer me up, telling me that I could come inside whenever I wanted. It was a consolation, of sorts.

Granny stuck her head in and handed Matthew the keys.

"Wait," I called to her as she started to walk away. "Why does he get the trailer?"

She turned to face me, her hands on her hips.

"He's the boy," Granny said, as if that settled it.

"But I'm older."

"Girls shouldn't sleep alone outside."

In the pause that fell between us, so much was said. She had to know this new sleeping arrangement left me vulnerable, but she remained silent, daring me to poke our family's secret.

"But what about me?"

"You have your own room now."

"But what about—"

Granny cut me off. "You can stay in our second bedroom if you have to," she said. "But don't make a habit of it."

Rather than admonishing Mom, holding a family meeting, or seeking professional counseling, instead of trying to figure out how to help Mom, Granny papered over the problem by getting Matthew and me panic rooms. Her solution tacitly reinforced Mom's behavior along with the idea that Matthew and I were the ones who needed to adapt to her unchecked moods. Mom couldn't cope with her own life, so Granny did it for her. My brother and I were remnants from a former life that Mom wanted to erase from memory. We were constant reminders of a future that was ripped away from her, our very existence making her feel an inexorable sense of failure. Granny's loyalty lay with her child; she would do whatever she could to soothe our mother and keep unpleasant realities away, even if that meant removing the unwanted burden of us from her.

I ducked back into the trailer and shut the door. I took a seat on one side of the dinette opposite Matthew. He had a dazed look of someone who had just lost something that was in their hand a second ago.

"You're so lucky," I said.

"I guess," he said.

"Did you ask for your own room?"

"No."

"Do you want to stay out here?"

Matthew shrugged. He was as perplexed as I was, but

just as powerless to change it. He pointed to a ledge that hung over the dinette.

"I can put a stereo up there," he said.

I was about to ask him where he was going to get a stereo when someone knocked on the door. Matthew opened it and Mom nudged him to one side and let herself in. With three inside it was like standing in a crowded elevator.

"Nice place you got," she said, turning in a circle for the full view. Then she reached for me. "Come here," she said sweetly.

She wrapped me in a warm hug. Despite my raw fear of her, I felt myself instinctively relaxing into her embrace. Her warm tears dropped on my shoulder. "I haven't been getting any sleep," she sniffed.

She released me and tilted my chin away from her to look at the fading scratch marks.

"Does it hurt?"

"Not anymore."

She looked out the open door and spoke with her head turned away from me.

"I love you, you know. But sometimes you make me so mad." I could hear her vigorously rubbing her stuffy nose. "I hate it when we fight. Let's not fight, okay?"

Her personality change was bewildering, but I went along with it to avoid any more trouble. "Okay," I said.

She hugged me one last time and stood to go. As she exited, Matthew and I watched to make sure she was leaving. She took a couple steps and then turned back around. She wore an impish smile.

"Hey," Mom called out. "Do you love me?"

I stood in the doorway and nodded.

"Oh yeah?" she said in a baby voice. "How much?"

This was one of our childhood games that we used to play in Rhode Island. She'd repeatedly ask how much I loved her, and I would answer, "This much," holding my hands farther apart with each response until they were as wide as they could be, my whole body in the shape of a T proclaiming my love.

I held my hands a foot apart. That much.

"How *muh-uch*?" she cooed, drawing the last word into a singsong of two syllables.

"This much!" I shouted, reaching my arms as wide as they would go. I felt like an actor playing me in a movie.

"Me, too!" Mom answered, beaming. And thus Mom decided everything was back to normal. As I watched her walk back to her house, I knew that I would never feel right about us again. Her house was not my home; it was a dangerous place where I needed to keep my wits about me, and a survival plan in place. Starting now, I would simply hang on and wait until I graduated high school and could make my escape. Meanwhile, I would go through the motions of being a daughter. I would stay out of her house as much as possible, and during the times we were together, I'd smile and feign pleasantries. If no one in the family would protect me from her, then I'd have to do it myself.

"That was weird," Matthew said.

"Indeed."

14

Bee Dance

1984–1986

My brother's camper marked our final pulling away from Mom, the turning point when we willfully went our separate ways. By the time I was fourteen, I had outgrown my hope that Mom would revive with a fresh start in a new home, accepting that it had been nothing more than an immature wish, as likely to become real as a child's prayer for a new pony. Her increasing volatility was never mentioned, but it was the unspoken catalyst for our grandparents to alter our living arrangements so that my brother and I could safely navigate around her.

Matthew and I gravitated back to the little red house to watch TV and do our homework, we ate dinner with our grandparents and afterward Matthew would steal away to his detached room while I lingered to play checkers or cribbage with Grandpa. I waited until dark, when I knew Mom would be settling down to bed, and crept back to my room at the opposite end of the rental house.

Our retreat drew no complaint or question from Mom, and we saw less and less of her, settling into our separate lanes of mutual avoidance with the relief that comes from no longer trying to force an unnatural relationship.

By the time Matthew was starting middle school and I was in my first year of high school, the three of us had the physical proximity of neighbors, along with the attendant emotional distance. It was a face-saving compromise that solved the immediate problem of our safety yet not the underlying one of our abandonment, but it worked because it avoided confrontation and gave the illusion that Mom was still our parent. With Granny's creative problem solving and Grandpa's silent acquiescence, Matthew and I were forced to accommodate a belief system that robbed us of our mother. It was like we lived with a functioning alcoholic and rather than speaking the truth, our family just kept filling her glass to keep her from antagonizing us.

Matthew, now twelve, had grown accustomed to living in a detached trailer. At first, he'd been afraid to sleep alone. He'd spent nearly his entire life sharing a room with Mom and me, and it had taken him about a week of tearful nighttime returns to the little red house before he got the hang of it. With the addition of lights and running water, courtesy of a hose and an extension cord, Matthew felt better, and now he spent most of his time sequestered inside. In summer, he left the windows and door open to circulate air, and in winter, when it got so cold in the trailer that he could see his own breath, he burrowed under several electric blankets. He'd decorated the walls with posters from the rock band Rush and installed a cheap stereo Granny picked up at an electronics store, transforming his lair into a thunderous, pulsating sound pod. He had formed a school rock band with a few friends, and he was forever tapping his drumsticks on something, insulated from Mom's outbursts and lost in a beat only he could hear.

He entered Mom's house only to use the bathroom and to change from pajamas to school clothes in the mornings in front of the plug-in heater. I made myself equally scarce, entering only to sleep, or for the occasional surreptitious meal when Matthew and I cooked macaroni and cheese or microwaved tacos in the kitchen when she wasn't home, being careful to clean and return everything to its spot afterward so we wouldn't provoke her.

When we did encounter Mom, our interactions had the forced courtesy of housemates bound by financial circumstances to share living space, but never went beyond a quick hello. She didn't ask questions about our lives, and we didn't inquire about hers. It was tacitly understood that Mom expected only occasional updates from us, and our grandparents could handle anything that we needed. As far as Mom was concerned, at twelve and fourteen, we were old enough to look after ourselves.

Granny stepped in to fill the void with busyness, packing our schedules with baseball and scouting, swimming lessons and art classes, and while all that activity insulated us from loneliness, it was another way to push our feelings to a far place where we couldn't access them, or even know that we should have them. We learned how to keep going, and to keep quiet.

Grandpa took Matthew and me with him to Big Sur every chance he could, and brought both of us into the honey bus during harvest season. As I got older, I detected a more serious undertone to his hive lessons—a gentle prodding to think beyond Via Contenta and to consider what we wanted, instead of what Mom needed. He spoke in metaphors, using the bees as examples of the proper way to behave. What he found noble and admirable in the way bees lived translated into his moral code for humankind, and in his subtle way, he encouraged us to embrace, rather than recede from, life. He reminded us that bees live for a purpose far grander than themselves, each of their small contributions combining to create collective strength. Rather than withdrawing from the daunting task of living, as our mother had done, honeybees make themselves essential through their generosity. By giving more than they took, bees ensured their survival and reached what might be considered a state of grace.

One summer morning Grandpa and I took the back way to his Big Sur hives, sloshing through Garrapata Creek and chugging over an abandoned logging road because he was tired of going the easy way through the eucalyptus and redwood groves of Palo Colorado Canyon Road. This off-road route was more exciting, because it was quite possible we could get the truck stuck in a ditch.

Branches of bay leaf and poison oak scraped our windows as he four-wheeled through the thicket, and poor Rita shot out from her bed beneath his seat and hopped into my lap.

I wrapped my arm around her trembling body and pulled her close. Our tires slipped on the dirt lane that was slick in places where spring water seeped out of the mountainside, and we bounced over a small rock slide that had tumbled down and scattered across our path. We managed to make it this time without getting stuck and having to call one of the Trotter brothers to rescue us with the winch.

As Grandpa got his gear out of the back of the truck, Rita and I headed for the creek to go hunt for tracks and scents left behind by animals. I was hoping to get lucky and find another keepsake, like that time I found a snakeskin.

When Grandpa was ready for my help, he whistled and the sound reverberated down the canyon. I stood up from some raccoon paw prints I was inspecting, and jogged back to the apiary. I put on the veil and Grandpa handed me the smoker. I sent a few puffs into the bottom entrance of the first hive, and the guard bees scurried back inside. Grandpa pried up the inner cover, and I heard the propolis seal give way with a sticky crack, exposing the ten hanging honeycomb frames inside the box.

The bees aligned themselves in rows in the open space between each honeycomb frame—each narrow gap precisely designed to be three-eighths of an inch to permit bee passage, but prevent the bees from building wax bridges and fuse the honeycomb sheets together. They poked just their heads above the top bars of the frames, to see who was breaking into their house. Their black heads all lined up looked like little shiny beans.

We waited a moment for the bees to adjust to the sudden loss of their roof. They stared at us cautiously, then a few brave ones broke ranks and crawled up to the top bars of the frames to swivel their antennae and assess the situation. It took only a second or two before they decided the threat was over, relayed the information to the other bees, and all of them began moving again, returning to work and ignoring Grandpa and me. Grandpa lifted the first honeycomb frame out, loaded down on both sides with bees, and gave it to me to hold so he could loosen the next frame.

By now I could hold a frame covered in bees and differentiate their individual job titles just by watching their behavior. I saw some housekeepers cleaning crystalized bits of honey from hexagon cells and receiver bees storing nectar in others, and builders repairing cracks in the wax comb. But my attention was drawn to one corner of the frame, where a single bee shook vigorously side to side, like it was being zapped with electricity. Its wings beat so fast they disappeared from view, and its body blurred to a black smudge. Then it suddenly stopped, as if catching its breath, took a few steps, then vibrated again. A group of bees had gathered to watch. I held the frame out to Grandpa and pointed.

"What's wrong with that one?"

"Nothing. There's your dancing bee."

Grandpa knelt down for a better look, and interpreted the dance for me.

"It's a field bee, and she found a really good food source, and she's telling the other bees how to find it," he said.

I watched the dancer walk in a straight line, making a sound I'd never heard from a bee before, a low rumble of a revving race car. She waggled her abdomen, then abruptly stopped, made a sharp right and circled back to her starting point, forming the capital letter *D*. Then she repeated her dance again. And again. Sometimes she turned left and made the *D* backward, but she always came back to the same starting place. Some of the bees cleared the floor for her, while others tripped behind her, trying to follow. She seemed possessed.

It wasn't how I'd pictured bee dancing. I thought bees danced together in a group, and more gracefully, maybe bopping up and down or swaying. This bee wheeled about the honeycomb in the grip of what appeared to be a full-blown tremor or crippling panic attack.

"What's she saying?"

Grandpa kept a small library of bee books, dating back to the 1800s, and had read the work of Karl von Frisch, a zoology professor who won the Nobel Prize for first deciphering bee dancing in Germany in 1944. Grandpa knew the dance steps were intentional, and conveyed three things—the direction, distance and quality of nectar and pollen. The angle of her wiggle walk, in relation to an imaginary straight line toward the top of the hive, was like an arrow pointing in the direction the bees should fly relative to the sun. How long she danced conveyed flight time from the hive, and the enthusiasm of her performance signaled the quality of the food. A passionate dance meant a really good discovery, maybe a swath of untapped sage coming into bloom.

Other field foragers take the directions and fly off to verify the dancer's information. If they like what they find, they will return to the hive and dance, too, passing the good news along to their hive mates.

As Grandpa was telling me all this, more bees had gathered for the performance, and soon the dancing bee had a small crowd. When she finally stopped shaking, her audience moved toward her to touch her.

"She sends a vibration while she dances, and the other bees hear it with their feet and know where to go," Grandpa said.

One by one, bees lifted into the air and headed west, deeper into the canyon to go find the treasure. I snapped my head up to meet Grandpa's eyes. He was grinning. I laughed out loud, pleased with this new wordless language he was teaching me.

I handed the frame back to him, and he slid it back into the hive.

"Can you guess what other type of bee dances?" he asked.

Right away I crossed off the lazy drones. Also the queen, who was too busy laying eggs to dance. Nurse bees don't leave the nursery to see what's outside, so they were unlikely candidates.

"Give up?"

I nodded.

"Scout bees."

I remembered Grandpa had explained to me that the scouts were the house hunters. When a growing colony is getting ready to divide, he said, they are the bees that select a new home and lead the swarm to it.

"Scouts dance to tell the bees where to relocate," he said.

Every spring, Grandpa put in overtime as a swarm catcher, so he knew a lot about them. When bee colonies outgrow their nesting space, they naturally divide themselves, with part of the colony flying off with the queen to rebuild a new colony somewhere else, and the rest staying behind to rear a new queen.

While a bee swarm looks like a disorganized frenzy in the air, Grandpa explained the event is actually planned in advance, with the bees discussing possible routes and withholding food from their queen to thin her for flight. The group must pick a warm day to depart and gorge themselves on honey first, so they won't die out in the cold while they are in between homes.

Initially, a swarm doesn't travel far from its original hive; they typically settle in a nearby tree or bush, and cluster there for a few hours or days until they make a group decision where to permanently nest. While hanging together, the swarm casts out hundreds of scout bees to house-hunt and come back to the group with options. Just like a forager dances on the honeycomb inside a hive to advertise flower patches, the scouts dance on top of the bee cluster to pass along the addresses of hollow trees, rock crevices or sometimes dry cavities within the walls of wood-frame houses as potential dwelling places.

Like people touring a list of open houses, the bees gather a list of addresses from various dancing scouts and go inspect their options. They fly into the advertised locations, taking measurements, checking the security of the entrance and feeling for drafts. They make their decision and return to the hive to dance with the scout whose home they prefer. As the energy and excitement builds, one scout reaches a tipping point of support, a consensus is reached and the entire swarm takes off with the queen to that scout's specific location.

The more I learned about bees, the more astounded I became with their social intelligence. Not only did bees have language, they were democratic. They researched, shared information, discussed options and made collective decisions, all for the betterment of the whole.

"You're right," I said.

"About what?"

"Bees *are* smart."

"You already knew that," he said.

"I didn't know they thought about the future."

Nothing about a bee colony was spontaneous; bees could see a problem coming and start making a change before it became serious and they perished. If their hive became overcrowded or unsafe, they took initiative to move to someplace better, abandoning a home that is too drafty or damp, too low to the ground where predators can get it, or too small for their growing family. Bees had enough brainpower to envision a better life, and then go out and get it. Even if it involved the risk of living out in the open, defenseless, until they decided together where to relocate. Bees had guts.

"What about you?" he asked.

Grandpa continued lifting frames one at a time out of the hive, examining both sides for eggs and larvae, and sliding them back in the box.

"What about me?"

"What do you see in your future?"

It felt like a trick question. "High school graduation," I said, which was three years away.

Grandpa put his hive tool into his back pocket, led me

a short distance from the hives and untied my bee veil. He tilted it off my head so he could look into my eyes. I could tell something was heavy on his mind.

"That's not what I mean," he said. "Have you thought about what you want to be someday?"

I realized with a sudden panic that I hadn't given it any thought. Grandpa was encouraging me to take a cue from the scout bees, and start planning now for my future. My grandparents' home was never intended to be more than a temporary stopgap, even though it had now been almost a decade. I couldn't live with him forever. And I could not live with Mom, ever. I was dangerously without a plan.

Grandpa was trying to tell me that I had to go out and find what I wanted, and then dance like hell for it.

"I'll go to college?" I offered.

"Now you're thinking," he said.

After our hive talk, I threw myself into high school. Every test, every essay, every science experiment was a chance to get a good grade, and the more A's I collected, the better the odds that a college would offer me a scholarship. I cared less about which school accepted me or even what I studied; I saw college more as a way to escape my living situation. The mere threat of spending the rest of my life on Via Contenta made me evangelical about homework.

I became a champion studier, turning in my book reports early to leave a good impression on my teachers. When I told Granny that colleges liked students with a lot of extracurricular activities, she posed as me and wrote a letter to the *Carmel Pinecone* offering to write a youth column for free. Not surprisingly, I got the job. Every two weeks

I typed a story on Granny's typewriter about high school events, she edited and fact-checked, and then I delivered the pages by hand to the editor in Carmel. A school counselor suggested athletics looked good on college applications, so I shifted sports teams with the seasons: diving, softball and field hockey. I lived in a whirlwind of my own making.

I dreamed of going to college but worried how to pay for it, so I got a job at the one place in Carmel Valley where teenagers could make decent tips—the local steakhouse. Will's Fargo was an old adobe roadhouse that Granny had lived in when she and her mother first landed in Carmel Valley in the thirties. It was a beloved favorite among the locals, kept in its original cowboy style, with a dimly lit saloon furnished with red velvet curtains, a fireplace, and mounted wild boar heads grimacing from the walls. Before sitting down, diners ordered their meals at a butcher station by pointing to the cut of meat they wanted, and the butcher sliced their steak, weighed it on a scale and pierced it with a wooden tag bearing the customer's name. The butcher slid the meat through a small door in the wall behind him, where the chef was waiting on the other side at the grill.

I was the dishwasher. I hosed dirty plates with a sprayer that dangled from the ceiling, then arranged them into a square plastic tray and slid it down a stainless-steel trough into a steaming industrial washer. It was equivalent to standing for eight hours in a sauna, not including the multiple trips hauling garbage bags to the dumpsters behind the restaurant. But I was thrilled to do it. I was a willing Sisyphus—no matter how many dishes I washed, the waiters came blowing through the swinging kitchen doors in their

Western vests and bow ties and dropped more plates into my sink. It was exhausting work that made the skin on my fingers peel, but the thought of college numbed the pain.

Waiters gave me a portion of their tips at the end of the night, supplementing my minimum wage paycheck. The money wasn't a lot, but the job also came with a huge perk. Before every shift, the chef fed the staff, and we could choose between steak, abalone or chicken, and the chef always made a soup and salad. I felt very grown-up, figuring out a way to feed myself and save for college at the same time. And the hours were ideal. I started at four in the afternoon and worked until midnight—ensuring Mom was fast asleep by the time I finished. I worked as many shifts as they wanted to give me.

As far as I was concerned, there was nothing I needed from my mother anymore.

Until I got my period.

I was almost fifteen and still had not had menstruation explained to me. I'd somehow missed out on any sex education, at home or at school, and aside from the bits of information I'd gleaned from friends about getting cramps and headaches, no one had prepared me for what to do when my time came. Although I wouldn't admit this to anyone, I didn't understand the physiology of where in my body the blood originated from, nor why. I knew in a vague way that it meant I was a woman and able to have babies, but that's where my knowledge stopped. I needed some type of feminine product, but I was unclear on what the different kinds were and which I needed. Granny seemed way too old to be able to help me with this one.

I found Mom in the living room, standing on a chair wearing three-inch Candies sandals identical to the ones Olivia Newton-John wore in *Grease*, misting her hanging spider ferns. She had just dyed her hair, and had a plastic bag stretched around her head and a towel with brown stains draped around her neck. She startled at the sight of me, and stopped in midspray.

"What's wrong?"

"I think I got my period."

"What do you mean, *you think*?"

"Well, I'm pretty sure."

"Is there blood?"

I nodded.

"Huh."

We stood there looking at each other, neither of us moving.

"Hang on," she said.

Mom stepped carefully off the chair and walked to her room, returning a moment later with her purse. She pawed through it and handed me a wadded-up five-dollar bill.

"Walk down to Jim's and get something for it."

She stepped back onto her chair and resumed watering her plants.

This wasn't how this was supposed to go. She was supposed to drive me to the store, show me what to get, have that mother-daughter moment when she tells me about her first time. I don't know...we were supposed to have *the talk* now, weren't we?

I was too mortified to buy sanitary supplies from old man Jim, a fixture behind the lone cash register who'd known me since I was small. He always rang people up slowly so he could inquire about new jobs and weddings and babies, or pass along the latest news of new jobs or weddings or babies. Jim knew the score of every Little League game, who got into college, who had recently died, and passed out cigars when a new baby came to Carmel Valley. He was our de facto town crier, and I was mortified at the thought of buying feminine products from the same man who still called me "kid" and slipped a candy bar into Granny's grocery bag when her head was turned.

I pleaded with Mom that it was too embarrassing. I would just *die* if Jim saw what I was buying.

"Nobody cares," she said with a wave of her hand. "Just go."

Mom lowered the stereo needle onto a Bee Gees record, and I watched her for a moment, humming along to "Night Fever" and spraying her plants, willing her to change her mind. Why couldn't she help me, just this once? The market was only a few blocks away, but I worried that by the time I got there, I might bleed through my jeans.

"Can't you just drive me?"

She pointed at the bag on her head and shrugged her shoulders, indicating she was in the middle of coloring her hair and couldn't leave the house. I folded the money into my pocket and returned to my room, where I tied a sweatshirt around my waist. I twisted open my piggy bank and took a few more bills from my college fund, then left through the screen door, slamming it as hard as I could on the way out.

"What's *your* problem?" she shouted after me.

At the market, I kept my eyes on the floor as I ambled toward the shelf with the feminine things. The pulse of teenage awkwardness knocked in my ears; I was terrified that someone may see me and learn that my body had become sexually mature. I was physically, but not mentally, ready for womanhood, and until I sorted that out, it was nobody's business but mine. I cursed Mom under my breath for staying home, then waited until the aisle emptied and quickly swiped a box of sanitary pads into my shopping basket. I chose the same brand I'd seen in Mom's bathroom, and quickly buried it under a box of Cheerios, a carton of milk and a loaf of bread. Just another kid on an errand for her mother.

Jim looked up from his crossword and smiled as I set the basket before him, and he rang me up as always, making sure to ask me how the bees were doing. He automatically plucked Mom's brand of cigarettes from the display behind him and asked me if she was running low. Matthew and I often made cigarette runs for her, but this time I wasn't sure I had enough money so I shook my head.

"Okay," he said, putting them back. "Grab yourself a candy bar, then."

Mom was in her room with the door closed when I got back. I set the grocery bag on the kitchen counter, took out the scandalous pads and scurried to the bathroom. I examined the packaging, read the directions and practiced walking with a pad between my legs. My transition to womanhood passed as quietly as a sigh. I didn't feel any different as I took my new woman-self back to my bedroom, and as I passed the kitchen, Mom was taking my groceries out of the bag with a puzzled expression.

"Did you buy all this stuff?"

I swallowed. I'd forgotten to hide my extra purchases in the cabinets so she wouldn't notice.

"I thought since I was at Jim's, I should also pick up a few things," I said.

I had never bought groceries before, so Mom stared at me a long time before she answered.

"That was thoughtful," she said. "But I guess you're right, as you get older you should start paying for some of the groceries around here."

I felt flattened. I should have known not to come to my mother for anything. She now saw me as an adult housemate who should divvy the grocery bill and not come to her with personal problems. I knew this, but it hurt each time I was reminded. When put in a situation that required her to put her needs aside for another person's, it did not compute. Her circuitry overloaded and she shut down. Her insatiable need to protect herself was not going to change, no matter how many times I hoped otherwise.

I did not say any of this. I smiled and told her it was a great idea for me to start buying my own food.

Then I left the house and returned to my grandparents' home, where I didn't have to pay for the right to exist.

By the time I became a junior, Granny had signed on as a volunteer in the high school career center, so she could get her hands on every college scholarship that came in and steer it my way before the other students had a chance to apply.

"It's not cheating, it's just being smart," she said. "Besides, you need the money more than those rich kids."

Mom didn't get involved in Granny's drive to get me to college, and while I was thankful someone was helping me, my grandmother's eagerness to plan this next step sometimes felt like it veered into a desire to get me out of her house. She gave me nearly weekly reminders that I had to bring home A's, because there was no way we could afford college unless I got a full scholarship. When my birthday came, she bought me luggage. She selected a handful of Bay Area colleges for me to apply to, corrected the grammar in my college essays and called the schools to check the status of my applications.

Our mailbox started filling up with college brochures, but the most persistent recruiter was Mills College, a private women's liberal arts school in Oakland. I didn't consider applying because it sounded like something out of *Pride and Prejudice*, but Granny announced she had signed us up for a tour.

We entered the campus through an impressive wrought-iron gate on a drive that took us through a row of ancient eucalyptus trees. We passed manicured lawns, and dorms built in the Spanish Colonial Revival style with stucco walls, terra-cotta tiled roofs, and balconies. The campus had bubbling fountains and a creek, an enormous library, and I learned Mills had chefs who prepare three meals a day for students, even baking the bread for toast. It looked more like a spa than a college.

But what impressed me the most was the students. I met a violinist, a rower, a ground squirrel researcher, a computer programmer and a fashion model—all in one day. They majored in mystifying things such as Political, Legal and Economic Analysis, or Sound Theory. These were women who didn't feel sorry for themselves, and I wanted to be near them so that I might absorb some of their confidence. By the time we left, I no longer cared that Mills was a single-sex campus. It was my first choice. They had an early-admission program, and I could apply immediately.

A few months after our visit, a student worker from the principal's office came into one of my classes and passed a note to my geometry teacher. He halted in the middle of his chalk equation and looked right at me.

"Meredith, can you come here, please?"

I went to the teacher's desk and opened the pink square of paper. "Call your granny," it read. I used a dime in the pay phone near the front steps of campus. Granny was out of breath when she picked up on the first ring.

"You got it!" she managed.

"Got what?"

"Mills sent you an acceptance letter. You got in!"

I opened my mouth, but no sound came out. My knees felt weak, and I gripped the edge of the metal box surrounding the pay phone to steady myself as colors melted and blurred around me. I could hear Granny catching her breath on the other end of the line. It was a win-win, signaling the beginning of Granny getting her life back and the start of mine.

"We did it!" she cheered.

Then I remembered the price tag. Thirteen thousand a

year. Private school tuition wasn't part of our family's vocabulary.

"But we can't afford it," I said.

"Don't worry, you are getting financial aid. We only have to come up with three thousand. Your grandfather and I will pay half, you and your mother can pay two hundred and fifty each, and you'll have to call your father to get the last thousand."

Granny had obviously been putting some thought into this. By patching together grants and loans from the school, the state and the federal government, and scraping together what our family could from selling honey, teaching and washing dishes, I was somehow going to go to college.

Walking back to class in the stillness of the empty hallway, I exhaled my first full breath of air in what felt like months, overwhelmed by the incredible notion that I now had a place to go. Relief felt like taking off smudgy glasses; the mundane became suddenly beautiful, and I saw new colors where I'd never seen them before: in the rows of scuffed brown lockers, the tramped-down crabgrass lawn where we ate lunch, in the crumbling adobe bricks worn concave inside the mortar of the school walls. Everything was just as it should be.

Although I had yet to hear back from schools I'd applied to in Berkeley, San Jose and Santa Cruz, I didn't want to wait. Mills was the first college to say yes, so I said yes, grabbing the first lifeline thrown to me. Like the bees, it was time to take a risk, go out there and choose a new home.

Later that afternoon, I knocked on Matthew's trailer door, loud enough so he could hear me over the pounding bass. He turned the music down and poked his head out of the door.

"You rang?" he baritoned, impersonating the butler Lurch from *The Addams Family.*

"Permission to enter, sir."

He swung the door open all the way, and took a step back so I had room to come in. He pushed aside a pile of music CDs on his bed to make space for me, and I sat down in a cross-legged position. My news fizzled out of me in one big whoosh.

Matthew clicked his stereo off and sat down next to me. "Wow."

I'd expected a slightly more celebratory reaction.

"That's it? Wow?"

He sat down next to me on the bed, put his elbows on his knees and his chin in his hands. "So that means you'll be leaving."

I was so self-centered. I'd been so focused on escaping that I hadn't considered what it would be like to be the one left behind. All this time I had been the natural buffer between Mom and Matthew, absorbing her hostility so he wouldn't have to. Now I was breaking an unspoken promise to keep him safe.

Mom had always aimed her neediness at me instead of him. Maybe because I was the first child, or because I was female, or perhaps because I looked so much like my father; I'd never know why she fixated on me and largely ignored my brother. She clung to me for comfort when we shared a bed after the divorce, while Matthew was banished to a small cot. She chased and cornered me, not Matthew, in the bowling alley. And even though we both consumed water and electricity, I was the one who took the punishment for the both of us.

Now, with a sick feeling, I worried that with me gone, Mom might finally focus on him.

"Be sure you keep staying out of her way," I said. "You'll be all right. She doesn't come out to the trailer."

"I know," he said.

He rearranged his face into a smile. "Hey, I'm really proud of you. I suppose now you're going to get all smart and stuff?" He swung open the door to his mini-fridge and held out a grape soda.

"Want one?"

I passed. He cracked it open, took a long glug and set it down in the sink.

"You know, she tried to hit me once," he said.

A pain shot up from my stomach to my temple, making me wince. "What?" I whispered.

I'd never seen her raise a hand to Matthew, and assumed he'd been spared.

"She took a swing at me, but I grabbed her arms and pinned her against the wall. I got right up in her face and told her never to touch me again, or she'd be sorry. I guess that scared her because she never tried it again."

Matthew was now taller and stronger than Mom. She probably sensed he could overpower her, so she backed down.

"Why was she mad at you?" I asked.

"I can't even remember what it was. You know Mom. Could've been anything. That's really not the point."

He picked up his drumsticks and began tapping out a rhythm on the wall.

I'd wished a thousand times that Mom had an understandable reason for off-loading us; I almost wanted her to have an addiction, something I could blame to remove the possibility that it was her choice. But she didn't drink. She never touched drugs. She didn't stay out late, leave us with strangers, nor did she bring men home. She was never institutionalized or homeless. She didn't gamble. She wasn't a religious zealot or a workaholic. She wasn't consumed by any of those things that can steal a mother and really screw up a kid.

Our mother simply wasn't.

"Why didn't you tell me?"

Matthew stopped drumming for a second.

"It was no big deal."

Not to me. This was a violation of our unspoken family rules. Matthew was supposed to be off-limits, but I had obviously failed to protect him. He had rescued me from her once, and I'd failed to do the same for him. Not only that, I was leaving him behind.

I tried to cheer us both up by reminding him that Oakland was only a couple hours away, and I'd be home for summers and holidays.

"And what about you?" I said, hearing Grandpa's words echo through mine. I imagined our grandfather was having similar conversations with Matthew about his future when they went to Big Sur.

"As soon as I'm old enough to drive, I'm gone," he said, slicing his hand through the air in an imaginary trajectory.

"Gone where?"

"Cal Poly, probably," he said.

Unlike me, Matthew already knew what he planned to study in college. A double major in music technology and graphic communication.

"Pick out another CD," he said, pointing with a drumstick at his pile of jewel cases.

I sifted through and handed him Dire Straits. "What do you think is wrong with her?" I asked.

The CD player slid out its tongue, accepted the disc and swallowed it again. Matthew paused with his finger over the play button.

"Seriously, Meredith? You'll never get an answer to that question."

Maybe he was right. But I had to give it one last try before I left her for good.

Although our relationship was permanently severed, I couldn't imagine, after all this time, simply walking away from her without an answer. I didn't want us to live the rest of our lives always wondering why we never could find a way to love each other. I needed to know what my family was hiding.

15

Spilled Sugar

1987

Mom was watching a Danish pastry spin in the microwave when I came into the kitchen one afternoon. She was wearing her pajamas all day again. I could hear an *I Love Lucy* rerun playing in her bedroom.

The microwave dinged. She reached inside, yelped in pain and dropped the steaming sweet to the floor. The swear words flew as she rushed to the sink to run her fingers under the tap.

"Mom!"

"Shouldn't be eating that on my diet anyway," she said.

I filled a dish towel with ice cubes and offered it to her.

"Thanks," she said, pressing it to her fingertips.

"Does it hurt?"

"Like a mother-scooter," she said.

I swiped the roll off the floor with a paper towel, then wet a second towel and rubbed the grease off the linoleum.

"You're a good kid," she said.

I could tell something was on her mind. Her show was beckoning, but she was lingering in the kitchen as if she wanted to tell me something. In these last few weeks before I left for college, we walked carefully around one another, not quite sure how to politely end our relationship. Both of us knew that soon there would be no artificial reason to keep us together anymore, beyond the perfunctory Christmas cards and birthday calls.

Mom poured herself some coffee that smelled like gingerbread cookies, and leaned against the counter, drinking it while still keeping her two burned fingers in the air. She looked at the ceiling as she spoke.

"So, I know I haven't been the best mother..."

Was this an overture? Did Mom want to make peace after all? She fidgeted with the amethyst ring Granny had given her, as I held my breath and waited for her to continue. She spooned more sugar into her mug and turned back to me.

"What I was going to say was, you know I did the best I could. At least you didn't starve."

True. She kept me alive. I had to give her that. But now that I was leaving, I'd been thinking about all the mother-daughter things we never did, and wondering if she had been doing the same. What would it have been like to go on a trip somewhere, to see her in the stands at my diving meets, or to just sit together in the house and talk about nothing special?

"All in all, I'd say you had it pretty darn good," she said, her voice brightening. "You certainly could've had it *a lot* worse."

She was filling in both sides of the conversation, what she wanted to say and how she wanted me to react. My job was to listen and agree, to make her feel better by replacing my reality with hers. I crumpled inside. This wasn't reconciliation; this was Mom wanting forgiveness for free.

"You think your childhood was hard. Mine was absolutely rotten."

Suddenly, she had my attention as her secret vault creaked open ever so slightly. She'd made many references to her ugly childhood over the years, but always brushed aside my questions, saying she didn't want to go into irrelevant history. But I'd never forgotten the one time we'd visited her father, how she'd left trembling in anger and it had taken her weeks to recover. She never told me why she was so upset with her father all those years ago. Now, maybe because our time was up, she was ready to talk. Even though I didn't drink coffee, I poured myself a cup and sat down, ready to listen.

"Tell me," I said gently. "What happened to you?"

She looked out the window toward Granny's house.

"My father was horrible to me, absolutely horrible."

She lowered her voice and spoke confidentially, as if she were ashamed of what she was about to say. She crossed her arms and grabbed hold of her shoulders, unconsciously protecting herself.

"Horrible how?" I asked.

"Every possible way you can think of."

Mom took a seat next to me, and with shaking hands pushed a piece of nicotine gum out of a small plastic tray and popped the tablet in her mouth. Apparently my brother's campaign to get her to quit smoking by taping magazine photos of blackened, cancerous lungs to the fridge was working. She chewed for a moment and scowled at the taste.

"Mom, tell me. What happened to you?"

She took a deep breath, and the words tumbled out of her.

"My father used to keep a long, thin tree branch, he called it his 'whipping stick,' on the mantel where I could see it," she said.

The first time her father hit her, she said, she must have been about three or four. Sometimes he used his bare hands, but he preferred his whip.

I flinched, as I pictured a horseback rider with a crop. Then I imagined a grown man, using the same instrument on a preschooler. I saw his hand rise up in slow motion, heard the whiz of the whip through the air and the piercing scream of a child. Mom had to be exaggerating; she couldn't have been that young. I asked her if she was certain she was remembering correctly.

"I'm sure," she said. "He'd make me go outside to choose the branch. I remember I wore these little red boots."

My face flushed with pointless vengeance. I couldn't go back and stop what had happened; I couldn't protect her from the rest of her own story.

"Oh, Mom."

Although her words were shocking, they had a ring of familiarity to them. It felt like I already knew Mom had been abused, but I'd never let myself truly believe it because it was too awful. It had been easier not to know. But I'd noticed little things; how she could barely stand to be in the same room with her father that one and only time we visited him. That Granny was so perturbed by her ex-husband that she couldn't bring herself to say his name, referring to him only as, "good-ole-what's-his-face." When I'd met this other grandfather, I'd had the uneasy feeling that I was about to be scolded. All I knew was that there was something dark and off-limits about him, something our family had purposely buried deep underground. But to ignore it was also to ignore Mom and the scars that remained inside her.

"How often did he hit you?"

Mom snorted derisively.

"Every couple of weeks? I don't know, so often I couldn't remember the reasons why anymore."

Mom spoke matter-of-factly, as if she were recounting the details of someone else's life, or a novel she had just read. Tears welled in my eyes at the thought of a grown man beating the innocence out of a little girl. But the thing that really cracked my heart was how casually she told her story, how she spoke as if it were an ordinary hardship not worth mentioning all this time. Time had dulled her outrage to the point that she nearly accepted the violence as her fate. But back when she was a little girl, how could she have ever understood that she had done nothing wrong? How can a kid make sense of adult rage?

I asked Mom why her father was so angry.

"There was no reason," Mom said.

Mom explained that her punishments weren't really in response to anything she'd done; her father beat her because he didn't like who she was.

"My own father despised me," she said. He told her she

was fat, that she was stupid. He hit her for being ugly. For moving too slowly.

"And you believed it..."

"I was just a child," she said.

"But you don't believe it anymore, right?"

Mom looked away without answering.

He'd trained her how to hate herself, blocking her from ever being able to love another person. Of course Mom was a bewildered parent. She'd never been shown unconditional love. So many things were starting to make sense. Mom's constant struggle with her weight, her crushing insecurity, her envious comments about how easily I made friends, how much I enjoyed high school. Why her divorce must have felt like her mythical glass slipper smashed to bits. Now I could understand why she chose to withdraw from a life she felt had cheated her at every turn. She had been groomed into victimhood; knocked down so many times it was safer to just stop trying.

She remembered her father lashing his leather belt on her for not clearing the table fast enough. After the whipping, she was sent back to finish collecting plates, but was so nervous that she dropped and shattered a porcelain sugar bowl.

"Then I got beat for the spilled sugar, too."

My breath caught in my throat. Mom was jumping from story to story now, as if telling tales at a dinner party. It wasn't sympathy or forgiveness she wanted from me, it was something much simpler. She wanted me to understand.

When she was five, she devised a method of escape. There was an oak tree in their front yard with long branches that grew low to the ground. One day Mom studied the tree and thought she might be able to dart up one of the long branches and disappear into the canopy if she got a good running start from the ground. So when her father was at work, she practiced, running and falling, running and falling off several different branches until she could do it. I pictured a plucky girl, like Scout in *To Kill A Mockingbird*, running barefoot in overalls, her hair messed up and scratches all over her skin, finally scampering successfully into the tree.

"Did you ever have to run into the tree?"

Mom chuckled.

"All the time. The first time I did it, he got so mad his face turned purple. I sure showed him!"

Mom was laughing now, relishing the one table scrap of childhood joy when she held power over him. I smiled with her, but it was forced. All these years, I hadn't known she carried this in her. Maybe if I had, I would have had more patience with her. Maybe if our family talked about the past, Mom could have healed. Instead we kept silent, and the abuse recycled through the generations. Her story spread like a spiderweb over the two of us, ensnaring us in its secrets.

I did a quick calculation. Mom hit me, her father hit her, so somebody must have hit him. I asked Mom what she knew about her father's childhood. Just the basics, she said: that his mother abandoned him when he was in elementary school, taking his sister with her. He was left behind with an alcoholic father who turned his fists on him.

Mom's beatings continued throughout middle and high

school, she said, only stopping when her parents divorced, not long before she left for college.

"Happiest day of my life was when he finally left."

It took me a second to absorb what she had just said. Her father hadn't crossed the line a handful of times. She had been traumatized her *entire* childhood.

"Where was Granny all this time?" I whispered.

Mom frowned.

"She knew what was going on but said nothing. I just hid the bruises and didn't talk about it. The one time I asked her why Daddy was so mad at me, she said that he wasn't a bad man, he was just tired."

I didn't know which was worse. The physical abuse or the mental torment of Granny gaslighting Mom into believing nothing was wrong.

"Granny never stood up for you?"

"She was afraid of him. He hit her, too."

I asked Mom how she could ever forgive Granny.

"She's my mother. She's all I have."

Her answer was profound, yet simplistic at the same time. Yes, we only get one mother. But are we required to forgive her? Where does a mother's needs stop and her child's begin? I told Mom I wasn't sure what I would have done in her shoes.

Times were different then, Mom explained, there was no such a thing as child protective services. Once, when her father hit her with a spatula and sliced open her thumb, Granny took her to the doctor's office and told the doctor exactly what had happened. He nodded knowingly and simply stitched Mom's thumb and sent them back home.

In a perverse way, the violence brought Mom and Granny closer later in life. They are survivors of the same war, Mom said, and eventually forgave each other for not thinking clearly when they were in the thick of it.

"Granny was trying to manage in her own way," Mom said. "She's certainly making up for it now. You should be thankful. If it weren't for her, we'd be living on the streets."

I could see why Granny took Mom in again and pampered her, trying to erase her guilt with a second chance at motherhood. They each overcompensated to fill a deep hole in the other, as if they were two broken humans who fused together into one whole person. Today they were emotionally inseparable. I always thought it was Mom who lacked the stamina to leave her mother's side, but now I could see how much Granny needed her to stay.

"Still, I wish Granny had protected you."

"Mom was in the house, but she just wasn't *there*," Mom said.

The echo of my own voice ricocheted around the room, mocking me. I had said the exact same words, countless times. Suddenly, my mother and I had something in common, and I felt a fleeting connection with her. We shared a similar suffering, one that maybe could be a starting place for us to try to understand one another.

I hoped that living apart would be good for Mom and me. We wouldn't be able to disappoint each other anymore. Maybe she could turn into the person she always felt Matthew and I prevented her from becoming. Maybe we still had a chance.

If there ever was a better moment for us to admit we

wished things had been different, it was now. I longed to tell her that I still hoped we could love each other one day. But after all the years of girding myself against her, the words felt like naive platitudes. I was too afraid to say them and have them not become true.

Instead I put my arm around Mom's shoulders and squeezed.

"Yes."

"Yes what?"

"You did the best that you could."

Mom sniffed, and dabbed at her eyes with the dish towel.

"Don't make the same mistakes I did. Go to college and get a job. Make sure you don't need a man before you marry one."

I gave her my word.

"Oh, and I almost forgot," she said, putting another pastry in the microwave. "I packed up some of your things you don't use anymore. You should go through the box and see what you want to take with you to Mills. What you don't want, I'll take to Goodwill."

Inside the box, I found my high school letterman's jacket, decorated with patches from the diving, field hockey and softball teams. I ran my fingers over the red brushed felt where my name was embroidered in cursive. My high school yearbooks were in the box, as well as my favorite quilt, my baseball mitt and cleats. Of course they were things I wouldn't use in college, but they were sentimental things I didn't want to give away to strangers, either.

Then, at the bottom of the box, my hand touched a book with a padded cloth cover. I recoiled, instantly recognizing my pink baby book. I had pored over the photos inside it as a little girl, trying to remember my forgotten family. By the time I was in second grade, I had memorized every page.

My skin went cold. Mom was not just clearing out my things; she was deleting all trace of me. A baby book was not something that went into the giveaway pile like an old coat. It was the one thing people grabbed when their house was burning down, the irreplaceable record of precious family history. The photos and memories written on these pages contained the only proof that Mom and I started out happy. I understood that Mom wanted to forget the past, but why couldn't she separate her children from the divorce? It was as if she, too, had been eagerly awaiting my college to begin, so she could finally be relieved of my constant reminder of her failed life. Ironically, she was throwing out the very thing that could have saved her. Matthew and I could have been her salvation, if she had allowed it.

I opened the cover. Inside was the mom that could have been. With a new mother's excitement, she had carefully documented each milestone of my first four years. She listed the dates when I first drank from a cup, the first time I smiled, my first step. There were photos of my first four birthdays, and details of my first trips—in a baby carriage, in a car to Boston, in a plane to visit Granny and Grandpa when I was one. Mom noted that I was doing well in YMCA swimming classes and that I liked school. When I first wrote my name, in shaky block letters, she taped it inside with an exclamation point–laden note about how I was developing ahead of my age group. Mom kept a running list of each new word I uttered, and recorded my first full sentence: "Where is Mommy?"

I turned the page and saw a wax envelope. Inside was a slippery lock of my brown baby hair, so many shades lighter than the almost black color it became. I shuddered at the thought of strangers pawing through my baby book at Goodwill, opening the envelope, touching my hair. A piece of my body that Mom threw away. Who would even want to buy a stranger's baby book?

I walked back into the living room and slipped the book back onto the bookshelf, where I hoped she wouldn't rediscover it. It seemed backward for me to be the guardian of my own baby book, and even more ridiculous to bring it to college. I wanted my mother to have it, like a normal mother, even if I had to trick her into keeping it.

I closed the box with my high school mementos and carried it outside. I could store it in Granny's house, where it would be safe from the Goodwill. Someday, when I was older, maybe even with kids of my own, I'd want to show them my yearbooks, or give them my baseball mitt and teach them to throw. But I left the baby book with Mom, more from stubbornness than anything else. Part of me was insisting she keep it, and part of me was testing her to see if she would.

I found Matthew in the driveway, bent over the open hood of a maroon Volkswagen Scirocco, tinkering with the engine. Matthew was also working at Will's Fargo restaurant, and had saved enough to buy the car. He taught himself how to change the oil and maintain the engine, and already had his driver's permit.

Matthew waved as I passed.

"What do you have in the box?" he asked.

I set it down and came over to see what he was doing.

"Do you know she wanted to give me back my baby book?" I said.

Matthew lowered the hood of his car and closed it with a thud.

"Follow me," he said, waving a greasy rag toward his trailer.

He reached into a cupboard over the sink and rummaged around. He pulled out his light blue baby book and tossed it over to me.

"She gave me mine, too."

Matthew started laughing, and then I did, too. The giggles swirled out of us, leaving tears and stomach spasms in their wake. I doubled over to try to stop, but it only made me laugh harder. We both flopped on his bed, grabbing our stomachs, unsuccessfully trying to shush each other. It was magically cathartic to share an inside joke with the only person on the planet who could truly relate. We'd both been dismissed, and therefore each of us could take it less personally.

Once our eruption settled, I opened his baby book. His was the same size as mine, but less than half the pages were filled. Born a year and a half before the divorce, Matthew had to compete for attention with a crumbling marriage. Mom's entries were factual and obligatory, lacking the detail and exclamation points of just two years before. Height. Weight. Date of birth. There is no travelogue of Matthew's first trips. Whereas Mom filled an entire page with each new word I spoke, she listed a mere handful for Matthew. After age two, Matthew's baby book went blank.

I handed him the book, and he put it back in the cupboard.

"Sorry to tell you, but you're not that special," he said.

Just then we heard the distinctive chug of the honey bus as it rumbled to life. Grandpa had a bumper summer crop now that the rains had returned, swelling the river and reviving the wildflowers.

"I'm going to miss that sound," I said. From the doorstep of Matthew's trailer, I could see into the bus. Grandpa was lifting the cheesecloth strainers off the top of his honey holding tanks. There were so many supers waiting to be harvested that he barely had space to maneuver.

"We should go help him," Matthew said.

Grandpa was standing on a milk crate peering into the barrels when we let ourselves in through the back door. He didn't hear us come in over the roar of the motor, and jumped when he saw us walking toward him. He hopped down from his perch and shut down the machinery.

"Barrels are full," he said, licking honey from his fingers. "You're just in time to help me jar. Gotta make more room before we can do another spin."

Matthew wriggled past Grandpa, and sat down before the tanks on the upturned milk crate and began filling jars with honey. Grandpa sidled next to him and lifted the gate on the spout of the neighboring drum. I took a position by the cardboard boxes of mason jars on the driver's seat, and handed them empties when they gave me their full ones. I screwed the lids on tight and stacked the honey jars on a plywood countertop resting over the uncapping trough. Sunlight filtered through the window and lit up the honey, casting spots of speckled amber light everywhere. It reminded me of the stained-glass windows in church.

The three of us moved together as if in a ballet, the honey passing from hand to hand, both Matthew and Grandpa so practiced that they could exchange their full jar with my empty one, and whisk it under the spout to catch the honey drip before it hit the floor.

This, I thought. This is what I will miss the most. That feeling of being exactly where I am supposed to be.

"You know," Grandpa said, breaking the silence, "I was forty when I married your grandmother."

He cleared his throat, and we waited for him to continue.

"So… I never thought I'd have kids."

I looked up from where I was now pressing Grandpa's honey labels onto a wet sponge and affixing them to the jars. Grandpa closed the honey gate and stood up, opened his arms wide and pulled us both in close to him. His voice dropped to a whisper.

"Then, lucky for me, you two showed up."

A feeling of joy burst inside me, carbonating through every pore. I did have a hive, and it was here, inside Grandpa's honey bus.

"I'll come home every summer, to help with the honey," I said.

"You'd better," Grandpa said, handing me another full jar.

Matthew looked up from his work.

"After I take my driver's test, maybe I can come up and

see you," he said. "We could go see a concert in San Francisco or something."

"Rush?" I suggested.

"What's Rush?" Grandpa said.

As Matthew tutored Grandpa on the genius of his favorite rock band, I dipped a finger into a jar of honey and brought it to my mouth. I tasted wild sage, salt from the sea and a nutty flavor like warm toast that ended with the faintest wisp of something sweet, like coconut. I felt the honey not only on my tongue, but also viscerally…in my memory and in my heart, as familiar to me as the sound of my own voice.

I could continue to define my life by all that it lacked, as my mother had done. Or I could be thankful that I had been rescued in the most profound way. Grandpa and his bees had guided me through a rudderless childhood, keeping me safe and teaching me how to be a good person. He showed me how bees are loyal and brave, how they cooperate and strive, all the things I'd need to be when it was my time to navigate solo. Grandpa had been quietly teaching me that family is a natural resource all around me.

Grandpa saw me helping myself to his honey.

"How much of that can you fit in your suitcase?" he asked.

"All of it," I teased.

Although I was leaving Grandpa's side, I would always feel his bees humming around me like an invisible force field, gently leading me to the right path.

They would protect me like they always had. Grandpa's hive lessons would never end.

EPILOGUE

2015

There is an old beekeeping myth that when a beekeeper dies, their bees mourn. The bees must be told that their caretaker is gone, otherwise they will become dispirited and lose their will to collect honey. They sense a disturbance in the order of things, which can cause them to despair and give up. The next of kin must shroud the hive with a dark cloth and sing to the bees to deliver the news and ask permission to become their new beekeeper.

One afternoon in 2015, Grandpa asked me to look after his bees. He made his request a month before he died.

He must have sensed that he was nearing the end. We had been sitting on his back deck watching his last remaining bee colony dart in and out of a sun-bleached pile of dilapidated hive boxes he'd tossed in a corner of the yard. He was eighty-nine, and no longer had the strength for beekeeping, but swarms kept finding their way into his abandoned equipment. He didn't inspect the bees anymore, but every afternoon he liked to sit in his deck chair and watch the foragers come home in the fading light.

His hand shook from Parkinson's disease as he pointed out the flight patterns. The bees were coming in from the south, from a patch of flowering ivy that had grown up the side of the neighbor's porch. They were feisty bees, he said, probably Russian stock, and hearty enough to make it through winter without any help from him.

"You'll take care of them for me?" he'd asked.

"Of course," I'd said, squeezing his hand and steadying the tremor.

I must have sensed a shift in Grandpa, too, because for the past several years I'd been making more of an effort to see him. I was forty-five, and had recently started a few hives of my own in San Francisco. I was finally making my way back to my grandfather, after way too long.

After graduating from college, I'd thrown all my energy into building a career in journalism, and had been so obsessed with chasing stories and switching newspaper jobs that I rarely got back home to Grandpa and the bees. I'd worked for six different Bay Area publications until finally making my way to the *San Francisco Chronicle.* I loved the symphony of ringing phones on the news desk, the lightning pace of a breaking story, and I kept a "go bag" with clothes, a toothbrush and maps in my trunk, ready to travel to a faraway assignment at a moment's notice. I was single-minded in my pursuit of a life that was constantly in transit and always on deadline.

But I felt my priorities shift as Grandpa started to decline. I stopped rushing around the globe, and spent my weekends sitting with him to watch the bees. Each time I visited, he gifted me another piece of his beekeeping equipment. I inherited his veils, his battered 1917 copy of *ABC and XYZ of Bee Culture*, the redwood jig he made to string wire in honeycomb frames.

By 2011, he'd cleared out most of his inventory and announced his reluctant retirement. It broke his heart to walk away from his bees after seventy years, and the bees must have felt the loss, as well.

But there was a way I could bring bees back to Grandpa. The same year, an editor and I installed two beehives on the roof of the *San Francisco Chronicle* building, convincing our bosses that it would be a unique way to report on the disappearing honeybee epidemic while trying our luck at urban beekeeping.

When the new bees arrived, I'd felt the vibration of their wings travel from my palms to my heart again, and I wept. I had not held bees in twenty-four years and their smell, their sound, their mannerisms were all so familiar—*so personal*—that I was overcome with a forgotten feeling of protection. My coworkers surely thought I was mad to cry over bugs, but how could I explain all that had passed between these tiny creatures and me?

Returning to beekeeping, I came to realize that I had a kid's knowledge of bees and needed to enlist Grandpa to mentor me in the finer aspects of colony nutrition, pest management, and especially swarm prevention because our hives were above one of the city's busiest intersections, crowded with bus stops, parking garages, bars and restaurants. I heard a new vibrancy return to Grandpa's voice as he advised me where to place the hives on the roof, or explained how to shake powdered sugar over the bees as a way to combat parasitic mites. We became a team again, and under his guidance, in four years I'd grown from a bumbling beekeeper into a halfway decent one.

That day in 2015 when he asked me to look after his bees turned out to be one of our last conversations. Not long after, he fell and broke his hip. Surgeons said it was inoperable, and five days later, Grandpa died.

I kept my promise to watch over his bees. That meant fetching his last hive and bringing it home with me.

Hives must be moved in the dark, when all the bees are huddled together inside keeping warm, otherwise some might get stranded in the field. I approached Grandpa's last hive in the predawn. I didn't have a funeral shroud, so I grabbed a dark blue dog towel from the back of my truck and draped it over the hive box. Then I tried to think of a song. I should have picked one in advance, because the Murphy's Law of singing is that when you try to recall lyrics they always elude you. Instead I knelt down next to the hive and placed my hand on the shroud, readying myself to just tell it straight to the bees.

To my left, there was an empty space where the honey bus once stood. A relative had torn it down for scrap, and the yard looked forlorn without it. It pierced my heart to see the deserted spot where it once stood, and I quickly looked away. I cleared my throat a few times, getting up my nerve to deliver the sad news to the bees.

"He's gone."

I waited, for what, I wasn't sure, some sort of sound or acknowledgment from the bees that they'd understood. I stayed crouched, listening in the early-morning quiet for a sign. A car started up somewhere in the neighborhood. A breeze rustled the leaves in the walnut trees. Life went on, just like it always did.

I lifted the shroud from the hive box, and still no bees came out. Maybe they weren't even inside anymore; maybe they'd died out or swarmed to a better location. Maybe those bees Grandpa and I liked to watch coming and going in the afternoon had only been robbers stealing abandoned honey or wax to use for their own hives. Maybe I just knocked on an empty house.

I took off the lid and peered inside with a flashlight. I saw four rotting frames, the honeycomb blackened with age and infested with white webs spun by wax moths. Ants ran amok, and a mouse had spent some time inside, judging by its paw-sized scrape marks in the honeycomb and the scat left behind.

But there was life, barely. There were about a thousand bees, a fifth the size of a starter package of new bees sold through the mail. The poor bees were trying to make a go of it, clinging to this small bit of rotted comb. They were pitiful, and clearly stressed out, zinging into my veil rapid-fire with a kamikaze anger I'd never seen before in a colony.

I leaned in closer. Bees pelted my veil like rain.

"It's okay, shhhh. You're okay."

I gently lifted a frame out and the colony practically shrieked. They were terrified, I'm sure, having never had their home invaded. Inside the hexagon cells I spied a miracle: white eggs. They had a queen. With a little care and feeding, this colony just might bounce back. I took out a second crumbling frame and carefully turned it to scan both sides until I found her—an all-black queen. She was the most striking matriarch I'd ever seen. Her abdomen lacked the usual stripes, each segment ink-black, and her thorax was dimpled by a single vertical line and ringed with a halo of yellow fuzz.

I transferred the three decaying frames of bees into a new hive box I'd brought with me. I centered the old honeycombs in the middle, between frames of fresh wax comb, so the colony had a clean place to put down honey and the queen had more room to lay eggs. I secured the lid with a ratchet strap, and duct-taped mesh screen over the hive entrance to keep the bees inside for the journey.

Grandpa's last wishes, which I had discovered typed on a sheet of yellowed paper buried in his sock drawer, requested that he be scattered at sea. I drove from Grandpa's house to meet Matthew at the Grimes Ranch in Big Sur, where Grandpa's cousin Singy unlocked the cattle gate to the pasture overlooking the Pacific. The emerging sun stage-lit miles of rocky coastline as my brother and I navigated quietly among the Herefords, admiring their deep red coats and white faces as they gnawed at the scrub, while being careful not to look the bull in the eye. We walked to the bluff, where seagulls extended their wings and soared in place, buffeted by the wind. I set down a wooden toolbox with a leather handle that Grandpa had made. Inside was a plastic bag from the mortuary containing his ashes.

We stood at the precipice of a twenty-foot drop where a thin tributary of Palo Colorado Creek cascaded into the sea. The waves barreled toward shore, smashing into the headlands and exploding through keyhole cliff arches carved by the sea. The ocean fizzed like seltzer shaken from a bottle, so angry that even the harbor seals had had enough and huddled on the few rocks remaining above the waterline, waiting for the ocean's mood to pass.

I opened the toolbox, untied the plastic bag inside and filled two of Grandpa's mason honey jars with his powdery ashes. Matthew let out a warbled sigh, and I wrapped my arms around his shoulders and clung so fiercely that I could feel the different rhythms of our heartbeats. It was just us now; we were a family of two. I wanted him to know in his bones that I would never leave him. The wind whipped our shirts and the sea howled as I whispered in his ear.

"I love you *so* much."

He sniffed, but didn't answer. I pulled apart slightly to look into his eyes, but he was staring at the ground. I tried again.

"You know that, right?"

Matthew looked at me for a brief moment, and then flicked his gaze downward once more. He nodded to let me know he'd heard, but also to move past my embarrassing outburst. Not his thing.

"Okay, on three?" he said.

We flung the ash from our jars in unison and the wind carried Grandpa into a dust comet over the waves. His particles hung there for a split second, and then vanished into the froth.

I suddenly remembered a conversation I'd had with Grandpa inside the honey bus when I was small. I'd asked him if he thought people went to heaven when they died.

"That's a buncha bull. You go into the ground and turn back into dirt," he'd said. It was a little shocking to find out that most adults had been lying to me; that there were no soft clouds, no angels with harps. Now, taking in the beauty of his final resting place, I appreciated how he had always been honest with me. I silently thanked him for giving me the respect of real answers.

Grandpa had returned to his ancestors. He was now part of these jagged mountains and that unruly sea. He was the pasture we were standing in and all the wildflowers in it, all the arrowheads buried under it, and every bee flying over it. He was the smell of wild Mexican sage blowing in the wind, and the cries of a baby sea otter that was bobbing on the waves and calling for its mother each time she dove in search of food. Grandpa was everywhere, so in one sense he was never gone.

Matthew and I waited for the mother sea otter to surface, to be sure she hadn't abandoned her pup, and then made our way back to my truck in silence.

I like to think Grandpa left on his own terms, much the same way a honeybee will abandon its hive to die alone if it is sick, to save the health of the colony. I believe that he didn't want to become a burden to his family, so he chose to remove himself as the ultimate act of sacrifice for the people he loved. The one saving grace was that Granny's heartbreak was softened by her dementia; she had a hard time remembering that her husband was gone.

Ten months later, she died in her sleep.

Mom's health plummeted after Granny passed away. Within a year, she was living in a care home where hospice nurses could monitor her adult-onset diabetes, and alleviate her chronic breathing problems with oxygen tubes. Each time Matthew and I visited, she seemed smaller, as if she were shrinking. When doctors used the term *actively dying* in the fall of 2017, Mom embraced the inevitable with a calm acceptance; seventy-three years of life had never been that good to her anyway, she reasoned.

Until the very end, I never knew what she was really thinking, if she was scared, if she had regrets, if she loved or despised me.

The last time she phoned me, she was exquisitely herself.

"I'm going to die soon," she'd said by way of opening, "and we've never had a good relationship. I want to know what you can say to me to make me feel better about that."

In her own way, I think she was expressing a need to make things right. She just wanted someone else to do it.

"We're okay, Mom," I said. "There's nothing to worry about anymore."

"You mean that?"

"Yes, Mom. Just rest."

I think I meant it. It was difficult to sort through the mixed emotions of losing the person who I'd wished all my life I could love. What kind of mourning is that, exactly? But the last thing I wanted to do was break my broken mother even further.

"I miss Granny," she said.

"I know, Mom. I know."

The last time I saw Mom was two weeks before she died. She was in a morphine haze, and Matthew and I were standing by her bedside, not certain if she knew we were there. Then all of a sudden her eyes flicked open and she seized my hand with the strength of a hawk's talons.

"I'm glad you're here," she mumbled before slipping back into sleep and letting go.

I was glad, too. I was glad that she could die knowing her children came to her, in the end. That she could feel some love in her life, even if it was so faint that it was sometimes hard to see. Ultimately, we are all social insects that thrive together or suffer alone.

When Grandpa asked me to watch over his bees, he didn't mean only his last colony; he was extracting a promise to care for all bees, for nature, for all beings. In short, he was asking me to see everything through the eyes of a beekeeper, to be gentle with all I encounter, even those things that can sting me.

I relocated Grandpa's last hive to a community garden in a postcard-perfect San Francisco neighborhood of pastel Victorians, where the streets were named after states and the air had a yeasty smell from the Anchor Steam Brewery.

It was an ideal honeybee sanctuary: a terraced urban farm at the dead end of a residential street. Behind a locked gate, there were two dozen individual plots and an elevated bee yard so the gardeners barely noticed the bees flying overhead. The hives received full sun, with heat radiating off a neighboring wall to provide ample warmth and a windbreak. All the bees had to do was exit the hive and descend directly over their private farmer's market, crowded with vegetables, citrus trees, lavender bushes and one beer maker's hop flowers. I imagine Grandpa would've approved.

Today, I think of him every time I open the hive, every time I harvest honey, every time I hear another apocalyptic news story about disappearing honeybees. I'm keeping my promise to him, and also repaying a debt by protecting the tiny creatures that protected me when I needed it most.

One morning, a class of preschoolers from the nearby bilingual school paid a visit to my apiary. The kids dressed in bright yellow safety vests and held hands as they walked together chittering in two languages about *abejas* and honeybees. They gathered around me under the shade of an apple tree, and once their teachers got them to stop wiggling, I knelt down to tell them a story.

"When I was not much bigger than you, I had lots of bees at my house," I said. "Bees are very special. Who can tell me why?"

"Because they make honeee!" one boy in a SpongeBob shirt called out.

"That's right! What else do bees do?"

Silence. The kids looked to one another for the answer.

"They fly?" said a girl with braids and a confetti of barrettes.

"They sting!" screeched another, reaching for the teacher's hand.

I was losing them. I stood up to show them the bee suit I was wearing, and pulled the veil over my head.

"I wear a special suit, so I'm safe. But bees are gentle. They won't bother you if you don't bother them. You don't need to be afraid."

I flipped the veil back to my shoulders and pointed to a raised garden bed. "What do you see growing in there?"

"Strawberries! Sunflowers! Cucumbers!"

"Would you believe the bees made those?"

I brushed my fingertip against a strawberry flower and showed them the yellow dust. "What's this yellow stuff?"

"Honey?" asked one boy.

"It's pollen," I said. "Flower dust. The bees get it on their feet and mix it up when they visit lots of flowers."

"In their pollen baskets!" chimed barrette-girl. Clearly, they had been studying bees in class. I was impressed.

"Exactly!" I said. "And when bees mix the pollen all around, it makes the flowers turn into food. Like a strawberry, or a cucumber, or a sunflower seed. Do you like to eat all those things?"

A fusillade of *yeses* fired into the air. Now they were primed for my main point.

"So the reason bees are so very, very special…is because they make our food!"

"They make honeee!" the boy in the SpongeBob shirt reminded me.

"Where's the queen?" a girl demanded, crossing her arms and jutting a hip. "I wanna see the queen."

I had no intention of opening the hive and risking someone's child getting stung. Or my queen getting squished by curious little folk. Now seemed like the right time to divert them with the honeycomb frame and let them poke their fingers in to taste.

The kids dug into the wax, dripping strings of honey into their mouths, giggling at the naughtiness of breaking something and making a mess. I felt a tug on my shirt, and saw a boy in cargo shorts and neon blue tennis shoes bouncing urgently, like he needed a bathroom. He was grinning maniacally, like he was colluding with me, but about what I wasn't sure.

I crouched down so he could have my full attention. He *really* needed to tell me something. The poor kid looked like he was about to burst.

"My grandpa has bees!" he shouted, jumping up and down like he'd just been given a puppy.

In that instant, all of San Francisco fell away and it was just this boy and me, alone in our private universe. We locked eyes and a shared thrill passed between us.

The little boy's eyes shone, and I saw the innocence Grandpa must have seen in me all those years ago. I wanted this boy to know that the world was big, so big that there were an infinite number of places to find love.

I knelt down, just like Grandpa used to do when he was about to tell me something important. I put my hands on the child's shoulders and whispered so that only he could hear.

"You are the luckiest boy in the *whole wide world*."

★ ★ ★ ★ ★

Acknowledgments

I am forever grateful to Heather Karpas at ICM Partners, who gave *The Honey Bus* its first green light. Her extraordinary talent, warmth and unwavering belief in this story kept me going, even when the road seemed long and twisting.

Editorial Director Erika Imranyi and the entire team at Park Row Books have done so much for me, and this book. It's been a dream collaboration; an adventure that never felt like work. Thank you, Erika, for your insightful touch—it's the secret ingredient on every page.

More applause for Helen Manders at Curtis Brown Group UK; and Maria Campbell and her whole team at Maria B. Campbell Associates, early champions whose endorsement ensured that the book would be translated around the globe. They have given Grandpa nothing less than immortality.

Group hug to my mentors David Lewis, reader of my

first draft, and Ken Conner, shepherd through the last. Both gentlemen were my editors at the *San Francisco Chronicle*, and I'm humbled and honored that they continue to guide me through words, and through life. Thank you also to friends who read parts of this memoir along the way and gave invaluable feedback: Earl Swift, Shobha Rao, Sarah Pollock, Meredith White, Julian Guthrie, Lysley Tenorio, Joshua Mohr, Tom Molanphy, Mag Donaldson, Tee Minot, Lesley Guth, Maria Willett, Maria Finn, and Maile Smith.

I am indebted to my creative nonfiction professors at Goucher College where this book began as my MFA thesis: authors Tom French, Diana Hume George, Leslie Rubinkowski, Laura Wexler, and Patsy Sims. A big bow of gratitude to the American Association of University Women, which supplied a generous grant so I could attend Goucher's graduate school program. This memoir was also supported by the Hedgebrook writers' residency on Whidbey Island, which, in a radical act of hospitality, loaned me a cabin in the woods to work on the manuscript.

This book was graced by all the beekeepers who opened their hives, their hearts and their homes: Noah Wilson-Rich in Boston; Aaron Yu, MaryEllen Kirkpatrick, Aerial Gilbert, and Deb Wandell in San Francisco; and in Big Sur: Peter and Ben Eichorn, Diana and Greg Vita; and my Nepenthe clan: Meredith, Kirk, and Will Gafill.

For their patience, understanding and generosity; love and thanks to my family. I wouldn't have had the strength to write this without the support of my brother, Matt; he protected me when we were young and countless times since. Thank you for being my confidante, for making me laugh and for making things right in the end.

To my father, David, who patiently answered my questions even when it was painful to do so, above all thank you for keeping the promise you made in 1975. You are, and always will be, my dad.

A never-ending thank-you to the honey in my life, Jenn. The bus seat next to me is forever saved for you.

Author's Note

I was fortunate to grow up in a place and time when honeybees were healthier, when I could walk into the bee yard certain I'd find life inside the hives.

But by and large, the world has turned against the honeybee since the days of the honey bus. Grandpa's prediction in the seventies of a widespread bee decline has come to pass, and the news is churning with apocalyptic stories of massive food shortages as we try to imagine a hungry planet without bees. I wish it were hyperbole, but when more than one-third of global crop production is dependent, wholly or in part, on bee pollination, it's hard to ignore.

What went wrong?

Honeybees thrived for fifty million years, but began to decline shortly after World War II, not long after farmers began using synthetic fertilizers instead of flowering clover and alfalfa cover crops to add nitrogen to their soil. Bee colonies took a dive from four and a half million colonies in the United States to just under three million today.

But commercial beekeepers in the United States first reported something was uniquely wrong in 2006, when they opened their hives after the winter frost expecting to see the usual—that most colonies had survived and about 15 percent had succumbed to cold or hunger and perished in a pile on the bottom board. But instead they found a mass exodus, anywhere from 30 to 90 percent of their colonies vanished from seemingly robust hives. Beekeepers had never seen anything like this before—healthy hives one day, and then ghost towns the next. Overnight, worker bees deserted hives brimming with honey and new generations in the nursery, leaving behind one stunned, unattended queen and only a smattering of hungry, lethargic newborns that had not yet learned to fly or feed themselves.

Funding poured into national labs, where entomologists raced to figure out what was happening. Emergency hearings were assembled as beekeeper after beekeeper told the same devastating story of sudden financial ruin. Beekeepers in Europe chimed in, saying their hives had been collapsing, too. In China, bee losses were so bad in some places that farmers had begun hiring people to pollinate crops by hand, spreading pollen to flowers with small paintbrushes.

This inexplicable catastrophe was given a clinical sounding name that implied the authority of a known cause but had none—Colony Collapse Disorder.

Scientists, beekeepers and activists have since offered a wide range of theories, blaming pesticides or fungicides, migratory beekeeping practices, the parasitic *Varroa destructor* mite, climate change, habitat loss, monoculture and various honeybee pathogens. While there is some promising research suggesting ways to boost honeybee immunity to withstand these threats, there's still no consensus on what's causing wholesale colony failure.

Europe has aggressively targeted neonicotinoids, a specific family of insecticides developed in the 1990s, commonly used to coat corn and soybean seeds before planting. Designed with a similar chemical structure to nicotine, the synthetic toxins are absorbed by the growing plant and affect the nervous systems of small insects, leading many researchers to conclude they disrupt a honeybee's ability to navigate back home. The European Union has experimented with a temporary two-year ban on neonicotinoids for flowering crops that attract bees, and a handful of states in the US no longer sell products containing neonicotinoids.

Results of these efforts are still being debated, with some pushing for a permanent ban and others arguing the experiments are inconclusive, wrongheaded, or worse for bees because they force farmers to switch to nonflowering crops or older, more toxic sprays.

Meanwhile, the bees are still struggling. While there has been a slight improvement in beehive survival since the shock of 2006, beekeepers continue to report to the United States Department of Agriculture that they are losing nearly one-third of their hives each year, a rate that is unsustainable over time even for a species that can multiply quickly.

Today, reports of colony collapse are just as inexplicably waning, and instead what a growing number of beekeepers say is killing their bees is not a mystery malady, but the parasitic *Varroa destructor* mite, a dark red creature no bigger than the head of a pin that attaches itself and sucks the body fluids of bee larvae and adult bees. They pass viruses to the bees that wreak havoc on the bee's ability to walk and fly, weaken its immune system and cause deformities such as wrinkled, useless wings.

Since *Varroa destructor* first appeared in the United States in 1987, the mites have continually developed resistance to the various organic and chemical methods designed to kill them. They can overtake a colony in days, multiplying exponentially each time a female mite enters a honeybee brood cell and lays eggs on the larva. The young mites are timed to hatch when the sickened bee emerges from the cell, unleashing a population explosion.

There's no easy answer to why bees are dying, but what's clear is that modern life has become increasingly stressful for honeybees, leading some in the beekeeping community to rename the epidemic Multiple Stressor Disorder.

I believe Grandpa was onto something when he predicted a man-made honeybee die-off. We are the ones who paved over the wildflower meadows. We are the ones who took bees out of their habitat and forced them to migrate on semis. We replaced small, diverse farms with monocrops, and then sprayed chemicals on the plants and trees we forced the bees to pollinate. The bees aren't to blame for overpopulation, factory farming, or lengthening droughts that dry out their flowers. But like the canary in the coal mine, they are dropping first. We have weakened the bees to the point they can no longer defend themselves against

Varroa destructor and a host of newer diseases such as *Nosema* gut pathogen and the Slow Bee Paralysis virus.

It's death by a thousand cuts for the bees. But what to do? People have to eat, so crops need to be pollinated. Birds, butterflies, bats, moths and ants pollinate, too, but can't possibly cover the millions of acres of produce the way honeybees can. Farmers need bees, but the paradox is that maybe we need them too much. We are squeezing the very lifeblood out of them to keep ourselves fed, and our farms profitable.

But we are also the ones who can use our ingenuity to help the honeybees live closer to the way nature intended. Lucky for us, bees are incredibly resilient and, if kept healthy, can propagate quickly. Across the globe, entomologists are working to breed hygienic, mite-resistant bees. Others are experimenting with mushroom teas to boost honeybee immunity. Citizen scientists are collecting hive data and helping keep track of bee populations. Gardeners are restoring landscapes with pollinator-friendly native plants. Farmers are switching to organic crops, and pushing the demand for nontoxic pesticides.

There's a growing consensus that we each have to do our own small act, whether it's seeding the roadsides with flowering plants, starting backyard hives of our own, or breaking up the food desert by planting flowering borders around mono-crops.

It's the principle of the hive—if each of us does our small part, it could add up to a bigger whole.

I owe Grandpa at least that much—to try.

And I owe it to the bees.

As long as honeybees stay strong, they can continue to pass their ancient wisdom to the next generation, so children can learn that even when they are overwhelmed with despair, nature has special ways to keep them safe.

My personality was shaped by the life lessons I learned in a bee yard. Every child should have that same opportunity to grow.

Questions for Discussion

1. *The Honey Bus* begins with a swarm-catching expedition gone wrong, and Grandpa has to rescue Meredith from stinging honeybees. Why do you think the book begins with this scene? How are the themes it sets up explored later in the story?

2. A major thread in *The Honey Bus* is the notion of biological versus chosen family. What kind of role do Grandpa and the bees play in Meredith's life, and how do they shape the person she becomes? Is there someone in your own life who had a similar impact on you?

3. Meredith's mother rarely leaves the bedroom and her mood sways between fragile and frantic. Grandpa, by contrast, is a soft-spoken Big Sur mountain man who loves the outdoors. How do these different personalities

affect the way Meredith sees the world? How do they dictate the family dynamics?

4. One way Meredith clings to the memory of her father is by listening to The Beatles, even though the music makes her cry. Does this resonate with your sense of music and visceral memory? Do you have songs that transport you back in time or make you feel strong emotions?

5. In reflecting on her childhood, Meredith writes, "I gravitated toward bees because I sensed that the hive held ancient wisdom to teach me the things that my parents could not. It is from the honeybee, a species that has been surviving for the last 100 million years, that I learned how to persevere." What does Meredith witness about honeybee behavior that informs her understanding of human nature and her own relationships? Has nature ever taught you something about yourself?

6. What was your comfort level with honeybees at the start of the book? Did it change by the end? How?

7. *The Honey Bus* title was taken from an actual hollowed-out, ramshackle army bus in the backyard where Grandpa bottled honey. When Grandpa teaches Meredith how to harvest for the first time, she writes, "The honey glowed in my hands, like a living, breathing thing. It was warm, and I loved it because it made sense when nothing else did." Throughout the story, Meredith and Grandpa keep retreating to the honey bus. What role does this space play in both of their lives?

8. When Meredith's brother, Matthew, is ten, he's given his own bedroom—in a camping trailer in the yard. Meredith envies his freedom, yet Matthew remembers shivering in the winters and feeling ostracized, sequestered outside until he eventually left for college. What do you make of this living arrangement, and how did it create different family experiences for the two siblings? If Matthew wrote a memoir, how do you imagine it would differ from his sister's?

9. In the epilogue, Meredith relocates Grandpa's last remaining beehive to San Francisco to start an apiary of her own in a community garden. A little boy visiting on a school trip tells her with pride that his grandfather keeps bees. Meredith tells him that he's "the luckiest boy in the world." What do you make of this final scene?

Further Reading

A Book of Bees, Sue Hubbell, 1988

ABC and XYZ of Bee Culture, A. I. Root, 1879

The Queen Must Die, William Longgood, 1985

The Honey Trail: In Pursuit of Liquid Gold and Vanishing Bees, Grace Pundyk, 2008

Letters from the Hive: An Intimate History of Bees, Honey, and Humankind, Stephen Buchmann and Banning Repplier, 2005

Honeybee Democracy, Thomas Seeley, 2010

The Life of the Bee, Maurice Maeterlinck, 1901

Langstroth's Hive and the Honey-Bee, L. L. Langstroth, 1853

The Bee: A Natural History, Noah Wilson-Rich, 2014

The Beekeeper's Lament, Hannah Nordhaus, 2011

The Beekeeper's Pupil, Sara George, 2002

New Observation's on the Natural History of Bees, François Huber, 1806

Field Guide to the Common Bees of California, Gretchen LeBuhn, University of California Press, 2013

Fifty Years Among the Bees, Dr. C. C. Miller, 1915

Bee, Rose-Lynn Fisher, 2010

The History of Bees, Maja Lunde, 2015

The Bees, Laline Paull, 2014

The Keeper of the Bees, Gene Stratton-Porter, 1925

Bees, A Honeyed History, Piotr Socha and Wojciech Grajkowski, 2015

Big Sur: Images of America, Jeff Norman and the Big Sur Historical Society, 2004

The Post Ranch: Looking Back at a Community of Family, Friends and Neighbors, Soaring Starkey, 2004

My Nepenthe: Bohemian Tales of Food, Family, and Big Sur, Romney Steele, 2009

These Are My Flowers: Raising a Family on the Big Sur Coast, The Letters of Nancy Hopkins, Heidi Hopkins, 2007

Recipes for Living in Big Sur, Pat Addleman, Judith Goodman & Mary Harrington, 1981

A Short History of Big Sur, Ronald Bostwick, 1970

The Esselen Indians of Big Sur Country: The Land and the People, Gary S. Breschini, 2004